A. S. Yakimov

Thermal Protection Modeling of Hypersonic Flying Apparatus

 Springer

A. S. Yakimov
Tomsk State University
Tomsk
Russia

ISSN 2520-8047   ISSN 2520-8055 (electronic)
Innovation and Discovery in Russian Science and Engineering
ISBN 978-3-319-78216-4   ISBN 978-3-319-78217-1   (eBook)
https://doi.org/10.1007/978-3-319-78217-1

Library of Congress Control Number: 2018935231

This Springer imprint is published by the registered company Springer International Publishing AG
part of Springer Nature
The registered company address is: Gewerbestrasse 11, 6330 Cham, Switzerland

Innovation and Discovery in Russian Science and Engineering

Series Editors

Carlos Brebbia, Wessex Institute of Technology, Southampton, UK
Jerome J. Connor, Department of Civil and Environmental Engineering,
Massachusetts Institute of Technology, Cambridge, MA, USA

This Series provides rapid dissemination of the most recent and advanced work in engineering, science, and technology originating within the foremost Russian Institutions, including the new Federal District Universities. It publishes outstanding, high-level pure and applied fields of science and all disciplines of engineering. All volumes in the Series are published in English and available to the international community. Whereas research into scientific problems and engineering challenges within Russia has, historically, developed along different lines than in Europe and North America. It has yielded similarly remarkable achievements utilizing different tools and methodologies than those used in the West. Availability of these contributions in English opens new research perspectives to members of the scientific and engineering community across the world and promotes dialogue at an international level around the important work of the Russian colleagues. The broad range of topics examined in the Series represent highly original research contributions and important technologic best practices developed in Russia and rigorously reviewed by peers across the international scientific community.

More information about this series at http://www.springer.com/series/15790

Preface

This monograph is devoted to studies of unsteady heat and mass exchange processes taking into account thermochemical destruction of thermal protective materials, research of transpiration cooling systems, thermal protection of composite materials exposed to low-energy disturbances, as well as a numerical solution of the conjugate problem of heat and mass exchange. The author proposes several mathematical models of passive and active thermal protection systems with regard to such complicating factors as ablation from the surface, surface roughness, phase transition of a liquid in porous materials, rotation of the body around its longitudinal axis, and exposure to low-energy disturbances. The author studies the possibilities to control thermochemical destruction and heat–mass exchange processes in transpiration cooling systems exposed to low-energy disturbances. The numerical analysis of the heat and mass exchange process in carbon plastics under repeated impulse action is given.

The numerical solutions of boundary problems are compared with the known experimental data.

The book is intended for specialists in the field of thermal protection and heat–mass exchange, as well as graduate and undergraduate students of physical and mathematical specialties.

Tomsk, Russia   A. S. Yakimov

Contents

1 Passive Thermal Protection 1

   1.1 Thermochemical Destruction of Carbon Phenolic
       Composites in the High-Enthalpy Pulsing Gas Flow 2

   1.2 Mathematical Modeling of Heat and Mass Transfer
       in Thermal Protective Coatings with Gas Flow Pulsation 13

   1.3 Mathematical Modeling of Surface Roughness and Ablation
       Effects on Thermal Protection 16

   1.4 Thermal Protection of Multilayer Structure Under
       Fire Exposure 21

2 Active Thermal Protection 31

   2.1 Mathematical Modeling of Heat and Mass Exchange
       Process in Heat Shielding Material 32

   2.2 Modeling of Heat and Mass Exchange Process in Transpiration
       Cooling System with Gas Flow Pulsation 42

   2.3 Modeling of Heat and Mass Transfer Process of Transpiration
       Cooling Systems Under Exposure to Small Energy
       Perturbations 49

   2.4 Modeling of Heat and Mass Transfer of Process of Systems
       Porous Cooling with Phase Transitions 51

   2.5 Modeling of Two-Phase Porous Cooling Process at Exposure
       Low-Energy Perturbations 60

3 Combined Thermal Protection 67

   3.1 Mathematical Modeling of Rotation on Conjugate Heat and Mass
       Transfer in High-Enthalpy Flow Around a Spherically Blunted
       Cone at an Angle of Attack 67

3.2   Numerical Analysis of Heat and Mass Transfer Characteristics
      in Radiative and Convective Heating of a Spherically
      Blunted Cone   81
3.3   Conclusion 101

Conclusion 103

Bibliography 105

Index 113

About the Author

A. S. Yakimov is a Doctor of Engineering Sciences, Senior Research Officer, Professor at the Department of Physical and Computational Mechanics, National Research Tomsk State University.

He graduated in 1970 from the Faculty of Mechanics and Mathematics, Kuibyshev Tomsk State University.

In 1981, he defended his thesis for the degree of Candidate of Physical and Mathematical Sciences at the Research Institute of Applied Mathematics and Mechanics, Tomsk State University (specialization 01.02.05 is Fluid, gas and plasma mechanics).

In 1999, he defended his thesis for the degree of Doctor of Engineering Sciences (specialization 05.13.16 is Application of computers, mathematical modeling and mathematical methods in research, and specialization 01.02.05) at Tomsk State University.

He is the co-author of a textbook, the author of one monograph and 84 publications (not including theses) devoted to mathematical modeling of some thermal protection issues, as well as to environment protection and mathematical solutions of equations in mathematical physics. The latter subject is addressed in the following publications:

1. Grishin AM, Zinchenko VI, Efimov KN, Subbotin AN, Yakimov AS. The Iterated Interpolation Method and Its Applications. Tomsk: Tomsk University Publishing House; 2004. 320 p.
2. Grishin AM, Golovanov AN, Zinchenko VI, Efimov KN, Yakimov AS. Mathematical and Physical Modeling of Thermal Protection. Tomsk: Tomsk University Publishing House; 2011. 357 p.
3. Yakimov AS. Analytical Solution Methods for Boundary Value Problems. Tomsk: Tomsk University Publishing House; 2011. 199 p.

4. Yakimov AS. Analytical Solution Methods for Boundary Value Problems. Tomsk: Tomsk University Publishing House; 2014. 214 p.
5. Yakimov AS. Mathematical modeling of thermal protection and some problems of heat and mass transfer. Tomsk: University Publishing House; 2015. 216 p.
6. Yakimov AS. Analytical Solution Methods for Boundary Value Problems. Boston: Academic Press is an imprint of Elsevier; 2016. 189 p.

Introduction

At the end of the last century and the beginning of the present millennium, much attention has been paid to developing and designing hypersonic flight vehicles both in Russia and around the world. When hypersonic flight vehicles enter the atmosphere of the Earth or of other planets at a cosmic velocity, their impressive kinetic energy [1] transforms into heat. This is why it is necessary to develop various thermal protection methods: ablative thermal protective coatings, forced feed of a coolant into the near-wall gas layer, reradiation of heat to the environment, etc.

With this in mind, a thermophysical calculation taking into account the capabilities of advanced materials should be regarded as an integral part of designing hypersonic flight vehicles.

In the 20th, mathematical modeling of heat and mass transfer processes became an essential method regarding thermal protection and the development of advanced structural materials for flight vehicles. In a number of cases, mathematical modeling is more cost-effective and often the only possible research method. This is largely associated with experimental conditions, such as high temperatures and pressures that arise when a flight vehicle enters the atmosphere at a high velocity.

Another important issue is to expand the operating temperature range. This makes it possible to approach real operation conditions of thermal protective materials when levels of convective heat flows from the gas phase reach $\sim 5.10^6$ W/m^2 or more.

Aerospace engineering uses different active and passive thermal protection methods associated with a variety of flight vehicle designs and their specific atmospheric flight conditions [2–19]. Passive thermal protection methods based on the application of ablative thermal protective coatings have been and remain the most popular solutions [2–15]. However, these methods have one disadvantage: They change the initial geometric shape of flight vehicles and, therefore, their aerodynamic characteristics in flight, which negatively affects the accuracy of ballistic parameters.

Numerous studies [15–25], including recent ones [2, 6, 10, 18, 19, 23, 26–31], suggest that active thermal protection systems of flight vehicles based on forced injection of a coolant into the boundary layer are very effective and promising.

An important advantage of these systems is that they do not change the geometrical shape and, therefore, the aerodynamic characteristics of flight vehicles, up to the end point of their flight path.

Under high heat loads, structural materials are often stressed to the limit of their performance capability. Development of combined thermal protection seems to be an alternative solution [25, 26, 28, 29, 32, 33]. Some studies [28, 29, 32] focus on the effects of heat-conductive materials that decrease surface temperature in the air heat curtain area. Materials with high thermal conductivity and injection of a coolant gas from the surface of the porous blunted area have been used to decrease the maximum surface temperature. As shown in [31], increased thermal conductivity of materials leads to a lower temperature of the thermal protective coating. Increased porosity ensures more uniform distribution of the coolant over the surface and decreases heat loads on the protected structure.

In real conditions, thermal protective materials are exposed to low-energy perturbations: acoustic vibrations, wall vibrations, and pulsations of gas flows [34–41]. Thermochemical destruction characteristics in such systems may vary widely. Stimulation of heat and mass transfer processes in continuous and permeable media is considered in [34–41]. Study [35] demonstrates that pulsating flows improve fuel–air mixing and reduce the extent of the combustion zone. The primary advantage of the method proposed in [35] is the high effectiveness of the process in terms of minimum pressure losses and maximum increase in temperature. As shown in [36], both stimulation and suppression of heat transfer are possible in the pulse impact jet as compared to a steady flow. An increased Reynolds number suppresses heat transfer, with all frequencies tending to a steady flow.

It is important to note that the energy cost of excitation is much less than the total energy of processes in the mechanics of reactive media [30, 31, 37–41].

In conclusion, let us briefly consider one of the trends in the development of promising thermal protection methods associated with body rotation. Rotation may ensure good stability in flight and, when combined with injection from the surface of thermal protective materials, change flow conditions and the body flow heat transfer [42]. In contrast to axially symmetric heating [32], for flow around the body at the angle of attack [27], the difference in heat flows on leeward and windward sides may be significant which leads to uneven heating. In order to reduce the influence of this effect, hypersonic aircraft imparts rotational movement around the longitudinal axis.

The possibility of attenuation of moderate-intensity laser radiation by combustion products of carbon-graphite materials was investigated [43].

Some results of numerical solution of boundary value problems are compared with known experimental data [2, 44].

In order to accomplish this objective, effective mathematical techniques and software systems need to be developed for solving spatial problems in the mechanics of reactive media. The proposed mathematical models and numerical solutions of spatial problems [45] are new. Using the method proposed in [45], some problems associated with control of heat and ablation conditions were solved for a wide variation range of Reynolds and Mach numbers, as well as angles of attack, with allowance for injection and thermochemical destruction.

This study extends the research into thermal protection and advanced composite materials that began in monograph [2] but not included in it.

Chapter 1
Passive Thermal Protection

Keywords Passive thermal protection · Composite polymeric materials
Non-deformable porous reacting body · Thermochemical destruction
Small energy disturbances · Heat and mass transfer · Carbon fiber
Surface roughness · Mass entrainment of gaseous filtration products
Coked layer

The development of aerospace engineering has given rise to the need to create lightweight high-temperature structures, polymeric and composite materials, fiberglasses and carbon fiber-reinforced polymers that are widely used as thermal protective materials [2, 4–14, 41, 44]. Polymer composites contain a thermoset plastic binder as a matrix (e.g., phenol-formaldehyde resins) and a filler-reinforcing elements in the form of glass and carbon-fiber fabrics. This polymer composite structure can be described as a porous body consisting of open and closed pores.

The interaction of polymer composites with high-enthalpy flows is a multistage physico-chemical process. This is accompanied by heating of the body, evaporation, pyrolysis of the binder, ignition, combustion, thermal destruction of the filler, dispersion of solid reagent, pyrolytic deposition of gas phase and heterogeneous reaction products on the surface of inflamed pores, and formation of carbon residues [4, 9].

Experimental research of each stage is associated with some technical difficulties, including inadequate equipment and measuring techniques, as well as the complexity of the process itself and the destruction mechanism of polymer composites.

Processes in heat power plants, flight vehicles, chemical reactors—wherever polymer composites are used as thermal protective coatings—are always accompanied by small perturbations generated by both external sources (wall vibrations, fluctuations, turbulent noises) and destruction of polymer composites in the gas flow containing an oxidizing agent [34, 41, 46]. Parameters of thermochemical destruction in such systems may vary widely.

The thermochemical destruction of carbon phenolic composites in the high-enthalpy pulsing gas flow [47] is modeled in Sect. 1.1 of this chapter. The heat and mass transfer process in carbon fiber-reinforced polymers with gas flow

© Springer International Publishing AG, part of Springer Nature 2018   1
A. S. Yakimov, *Thermal Protection Modeling of Hypersonic Flying Apparatus*,
Innovation and Discovery in Russian Science and Engineering,
https://doi.org/10.1007/978-3-319-78217-1_1

pulsations [48] is modeled in Sect. 1.2. Some effects of surface roughness and ablation on thermal protection [49] are studied in Sect. 1.3. The thermal protection of multilayer materials under fire exposure [50] is modeled in Sect. 1.4.

1.1 Thermochemical Destruction of Carbon Phenolic Composites in the High-Enthalpy Pulsing Gas Flow

Stimulation of heat and mass transfer processes in continuous and permeable media has been considered in a number of studies [34–41, 44, 46–48, 51, 52]. Study [46] identifies the limiting conditions for binder concentration and intensity of oscillations when it is possible to reduce heat loads on the wall and to control heat and mass transfer [46]. Study [37] reveals one of mechanisms for stimulating transfer processes in a liquid for the pulsating flow case (redistribution of gradients in the flow). Paper [38] focuses on sound field effects on stimulation of heat and mass transfers in the boundary layer. Paper [39] studies the effectiveness of the catalytic process in case of forced external effects and uses the non-stationary methods for generating catalytic processes to stimulate heat and mass transfer. Paper [51] studies the combustion of carbon-fiber-reinforced polymers in the low-temperature plasma jet with transverse harmonic vibrations of the surface. The expression for additional heat transfer in the porous body under periodic pulsations of a coolant gas was developed in study [52].

The mathematical modeling of thermochemical destruction for polymer materials based on carbon phenolic composites is discussed in [4, 9, 13, 14]. For the purpose of this section, a model of an undeformable porous reacting medium is used [9].

Problem Statement. The analysis of thermogravimetric measurements [53] shows that the pyrolysis of coal-plastic based on a thermoreactive polymer binder is a multiphase process. It includes such phases as decomposition of a polymer binder with heat absorption, generation of an intermediate condensed product (pyrosol), and a final condensed product (coke). The coke generation phase can be interpreted as a synthesis reaction of exothermic type [53].

The physics of the process in the condensed phase is as follows: [9, 13]. Under the effect of the high-temperature flow, temperature of coal-plastic increases to the decomposition point of a binder (resin). Then, pyrolysis of a thermoreactive binder begins, with formation of pyrosol and carbon residue (coke) which is retained within the matrix of reinforcing carbon fibers.

$$\nu_1 A_1 + \nu_4 A_4 \rightarrow \nu_2' A_2 + \nu_4' A_4 + \nu_5' A_5 \rightarrow \nu_3'' A_3 + \nu_4'' A_4 + \nu_5'' A_5, \qquad (1.1.1)$$

where $v_1, v_4, v_2', v_4', \ v_3'', v_4'', v_5', v_5''$ are the stoichiometric coefficients, $A_i, i = 1, \ldots, 5$ are, respectively, the symbols of the initial condensed binder, intermediate condensed product (pyrosol), final condensed product (coke), carbon fiber armor, and gaseous product of pyrolysis reaction. At $T_w > 800$ K, carbon surface is destroyed by the reaction with dissociated air components [9, 13, 54]. Gaseous products of the pyrolysis reaction can be filtered to the interface of media $y = 0$, blown into the boundary layer and, along with carbon oxidation products, reduce the convective heat flow coming to the body.

We make the following assumptions for the problem statement:

1. the Reynolds number in the incident hypersonic airflow is sufficiently high ($Re_\infty \gg 1$), and the boundary layer has formed in the vicinity of the body surface;
2. air at the outer edge of the boundary layer is in thermochemical equilibrium and composed of five components: O, O_2, N, N_2, NO;
3. the transfer in the boundary layer is considered on the basis of the simplifying assumptions that diffusion coefficients are equal to each other and Lewis number $Le = 1$;
4. in order to calculate the composition at the interface of gas and condensed phases, we will use the analogy of heat and mass transfer processes [4] on the assumption of "frozen" chemical reactions inside the boundary layer;
5. the following heterogeneous reactions proceed on the external surface of the thermal protective material at $T_w \leq 2400$ K:

$$C + O_2 \rightarrow CO_2, 2C + O_2 \rightarrow 2CO, C + O \rightarrow CO, C + CO_2$$
$$\rightarrow 2CO, O + O + C \rightarrow O_2 + C, N + N + C \rightarrow N_2 + C \qquad (1.1.2)$$

The findings of the study [54] can be used for a higher temperature range $T_w > 2400$. This study reveals the key features of the carbon ablation mechanism based on the complete thermochemical model.

In order to calculate the composition at the interface of gas and condensed phases, we will use the analogy of heat and mass transfer processes [4] on the assumption of "frozen" chemical reactions inside the boundary layer [13]. In case of increasing the stagnation pressure, the model of a chemical equilibrium boundary layer is more consistent with real conditions. However, for composites with high catalytic activity in relation to dissociated air components, heat flows in both cases do not differ very significantly. The numerical calculations of surface destruction [55] show that this approach can be applied to estimate the ablation rate.

Let us consider the chemical kinetics of heterogeneous processes on the body surface. If the serial number of a component corresponds to the following order of enumeration: O, O_2, N, N_2, CO, CO_2, molar velocities of reactions (1.1.2) take the form [9, 13]:

$$U_1 = \frac{k_{1w}\rho_w c_{2w}}{m_2}\exp\left(-\frac{E_{1w}}{RT_w}\right), \quad U_3 = \frac{k_{3w}\rho_w c_{1w}}{m_1}\exp\left(-\frac{E_{3w}}{RT_w}\right),$$

$$U_2 = \frac{k_{2w}\rho_w c_{2w}}{m_2}\exp\left(-\frac{E_{2w}}{RT_w}\right), \quad U_4 = \frac{k_{4w}\rho_w c_{6w}}{m_6}\exp\left(-\frac{E_{4w}}{RT_w}\right) \qquad (1.1.3)$$

$$U_5 = \frac{k_{5w}\rho_w c_{1w}}{m_1}, \quad U_6 = \frac{k_{6w}\rho_w c_{3w}}{m_3}$$

By using (1.1.3), we will find the mass rates of generation (disappearance) of components as a result of heterogeneous reactions:

$$R_{1w} = -m_1(U_3 + U_5), \quad R_{2w} = -m_2(U_1 + U_2 - U_5/2), \quad R_{3w} = -m_3 U_6,$$
$$R_{4w} = m_4 U_6/2, \quad R_{5w} = m_5(2U_2 + U_3 + 2U_4), \quad R_{6w} = m_6(U_1 - U_4)$$

and the expressions for the ablation rate:

$$(\rho v)_{2w} = (\varphi_4\rho)_w\left[\left(\frac{m_6}{m_2} - 1\right)c_{2w}B_1 + \left(2\frac{m_5}{m_2} - 1\right)c_{2w}B_2 \right.$$
$$\left. + \left(\frac{m_5}{m_1} - 1\right)c_{1w}B_3 + \left(2\frac{m_5}{m_6} - 1\right)c_{6w}B_4\right] \qquad (1.1.4)$$

$$m_e^{-1} = \sum_{\alpha=1}^{6}\frac{c_{\alpha e}}{m_\alpha}, \quad B_i = k_{iw}\exp\left(-\frac{E_{iw}}{RT_w}\right), \quad i = \overline{1,4}, \quad \rho_w = \frac{P_e m_w}{RT_w}, \quad m_w^{-1} = \sum_{\alpha=1}^{6}\frac{c_{\alpha w}}{m_\alpha},$$

where c_{iw}, $i = 1, 2, \ldots, 6$ are the mass concentrations of the components at the interface of gas and condensed phases [9]; m is the molecular weight; ρ is the density; k_{iw}, $i = 1, 2, \ldots, 6$ is the before-exponential factor; E_{iw}, $i = 1, 2, 3$ is the activation energy; T is the temperature; P is the gas pressure in pores; R is the universal gas constant.

The balance relationships for mass concentrations of the components (c_{iw}) can be written using Fick's law for diffusion flows, as well the analogy of heat and mass transfer processes [4, 9]:

$$J_{iw} + (\rho v)_w c_{iw} = R_{iw}, \quad i = \overline{1,6},$$
$$J_{iw} = \beta(c_{iw} - c_{ie}), \quad \beta = \alpha/c_p, \qquad (1.1.5)$$

where the total ablation rate $(\rho v)_w = (\rho v)_{1w} + (\rho v)_{2w}$, $(\rho v)_{1w}$ is the ablation rate associated with pyrolysis; $(\rho v)_{2w}$ is the ablation rate associated with heterogeneous chemical reactions (1.1.2).

The disintegration products at $T_w < 2400$ are assumed to weakly dilute the air mixture in the boundary layer. This makes it possible to use the above-listed assumptions (1)–(5) in the boundary layer.

The mathematical formulation of the problem for the condensed phase is based on the following assumptions:

1. for the sake of simplicity, the coal-plastic thermochemical destruction process is one-dimensional;
2. the filtered gas is uniform, with the molecular mass being close to that of the air mixture;
3. the porous medium is ideal and assumed to be one-temperature in the heat transfer process;
4. the gas flow inside flared pores is laminar and described by the linear Darcy law [4, 9].

The first assumption is sufficiently accurate since the size of the reaction zone is on the order of 2.5×10^{-3} m at $T \geq 600$ K; the heated layer thickness is less than 4×10^{-3} m (see Fig. 1.3) and considerably less than the original thickness of the thermal protective material: $L_0 = 10^{-2}$ m.

The second assumption is taken for simplicity of the model since the chemical composition of filtering gas products generated from thermochemical destruction [46] is not known for the used carbon phenolic material (P5–13N).

In the case of a two-temperature medium under destruction [56] (in contrast to the inert medium [4]), the volumetric heat transfer coefficient A_v is usually determined with a significant error. According to [56], the difference in temperature of solid and gas phases at the most real values $A_v = 10^6$ to 10^8 W/(K m^3) does not exceed 323–373 K, provided that the temperature of the coke residue is 1273–2273 K.

Finally, the real thermochemical destruction of coal-plastic [46] is accompanied by transient and turbulent gas filtration in pores described by the nonlinear Darcy law [4, 9]. However, in the case of a chemically reacting medium, viscosity and inertial coefficients for the material considered below [46] have not been found in the available literature.

The varying (pulsating) convective heat flow $q_w(v, t)$ is assumed to act on the thermal protective material for a definite time

$$
\begin{aligned}
q_w &= \frac{\alpha}{c_p}\left[1 - \frac{k(\rho v)_w}{\alpha/c_p}\right](h_e - h_w), \\
\frac{\alpha}{c_p} &= \left(\frac{\alpha}{c_p}\right)_0\left[1 + \frac{A \cos(2\pi v t)}{(\alpha/c_p)_0}\right],
\end{aligned}
\tag{1.1.6}
$$

where $(\alpha/c_p)_0$ is the heat transfer coefficient without pulsations; A, v are, respectively, the amplitude and frequency of pulsations; t is the time; h is the enthalpy; k is the attenuation coefficient for turbulent flow in the boundary layer.

Note that according to the experimental data [46], the turbulent flow around the thermal protective material was observed in the boundary layer. For this case, the attenuation coefficient $k = 0.19 \cdot (m_e/m_w)^{0.35}$ in the heat flow (1.1.6) was taken in the form of V. Mugalev's modification [4].

Mathematically, the problem can be reduced to the system of equations written in the moving coordinate system [9, 47] associated with the thermochemical destruction front:

$$c_p\left(\frac{\partial T}{\partial t} - \omega\frac{\partial T}{\partial y}\right) + c_{p5}\rho_5\varphi_5 v\frac{\partial T}{\partial y} = \frac{\partial}{\partial y}\left(\lambda\frac{\partial T}{\partial y}\right) - q_1 R_1 + q_2 R_2 \qquad (1.1.7)$$

$$\frac{\partial \rho_5\varphi_5}{\partial t} - \omega\frac{\partial \rho_5\varphi_5}{\partial y} + \frac{\partial \rho_5\varphi_5 v}{\partial y} = (1-\alpha_1)R_1 + (1-\alpha_2)R_2, \qquad (1.1.8)$$

$$\rho_1\left(\frac{\partial \varphi_1}{\partial t} - \omega\frac{\partial \varphi_1}{\partial y}\right) = -k_1\rho_1\varphi_1\exp\left(-\frac{E_1}{RT}\right), \qquad (1.1.9)$$

$$\rho_2\left(\frac{\partial \varphi_2}{\partial t} - \omega\frac{\partial \varphi_2}{\partial y}\right) = \alpha_1 k_1\rho_1\varphi_1\exp\left(-\frac{E_1}{RT}\right) - k_2\rho_2\varphi_2\exp\left(-\frac{E_2}{RT}\right), \quad (1.1.10)$$

$$\rho_3\left(\frac{\partial \varphi_3}{\partial t} - \omega\frac{\partial \varphi_3}{\partial y}\right) = \alpha_2 k_2\rho_2\varphi_2\exp\left(-\frac{E_2}{RT}\right), \qquad (1.1.11)$$

$$\rho_4\left(\frac{\partial \varphi_4}{\partial t} - \omega\frac{\partial \varphi_4}{\partial y}\right) = 0, \qquad (1.1.12)$$

$$P = \frac{\rho_5 RT}{M_5}, \quad v = -\frac{z}{\mu}\frac{\partial P}{\partial y}, \quad z = \frac{z_*\varphi_5^3}{(1-\varphi_5)^2}, \qquad (1.1.13)$$

$$z_* = d_p^2/120, \quad \varphi_5 = 1 - \sum_{i=1}^{4}\varphi_i, \quad c_{p5} = b_1 + 2b_2 T,$$

$$\lambda_5 = \lambda_{5*}\sqrt{\frac{T}{T_*}}, \quad \mu = \mu_*\sqrt{\frac{T}{T_*}}, \quad (\rho v)_{1w} = (\rho_5 v\varphi_5)_w,$$

$$\omega = \frac{(\rho v)_{2w}}{(\rho_1\varphi_1 + \rho_2\varphi_2 + \rho_3\varphi_3 + \rho_4\varphi_4)_w}, \quad \lambda = \sum_{i=1}^{5}\lambda_i\varphi_i \qquad (1.1.14)$$

$$c_p = \sum_{i=1}^{4}c_{pi}\rho_i\varphi_i, \quad \alpha_1 = \frac{v_2'M_{2s}}{v_1 M_{1s}}, \quad \alpha_2 = \frac{v_3''M_{3s}}{v_2'M_{2s}}$$

$$R_1 = k_1\rho_1\varphi_1\exp\left(-\frac{E_1}{RT}\right), \quad R_2 = k_2\rho_2\varphi_2\exp\left(-\frac{E_2}{RT}\right)$$

The system of Eqs. (1.1.7)–(1.1.12) should be solved by taking into account the following initial and boundary conditions:

$$T|_{t=0} = T_0, \rho_5|_{t=0} = \rho_{5,0}, \quad \varphi_i|_{t=0} = \varphi_{i,0}, \quad i = 1,\ldots,4; \qquad (1.1.15)$$

$$q_w - (\rho v)_{1w}(h_w - h_g) - (\rho v)_{2w}(h_w - h_c) - \varphi_4 \varepsilon \sigma T_w^4 = -\lambda \frac{\partial T}{\partial y}\Big|_{y=0-s(t)}, \quad (1.1.16)$$

$$P_e = P_w\big|_{y=0-s(t)}, \quad P_e = \rho_e T_e R \sum_{i=1}^{5} \frac{c_{ei}}{m_{ei}}, \quad (1.1.17)$$

$$T\big|_{y=l} = T_0, \quad v\big|_{y=l} = 0, \varphi_i\big|_{y=l} = \varphi_{i,0}, \quad i = 1, \ldots, 4, \quad (1.1.18)$$

$$l = L_0 - s(t), s(t) = \int_0^t \omega(\tau)d\tau, h_w = \sum_{\alpha=1}^{6} h_\alpha c_{\alpha w}, \quad h_g = b_1 T_w + b_2 T_w^2 \quad (1.1.19)$$

Hereinafter, b_i, $i = 1, 2$ are the constants; c_p is the heat capacity coefficient at constant pressure; L_0 is the initial thickness of coal-plastic; l is the variable thickness of coal-plastic; M_5 is the molecular weight of gaseous pyrolysis products; d_p is the diameter of the tubular pore; k_i, $i = 1, 2$ is the before-exponential factor for the binder and pyrosol decomposition reaction; E_i, $i = 1, 2$ is the activation energy of the last reaction; q_i, $i = 1, 2$ are the thermal effects of this reaction; $s(t)$ is the burn-up depth; v is the rate of gaseous products in the binder decomposition reaction; z is the permeability coefficient of coal-plastic; α_1, α_2 are the reduced stoichiometric coefficients [9] from the pyrolysis kinetic scheme (1.1.1); β is the mass transfer coefficient; ε is the integral thermal emissivity; $\varphi_i, i = \overline{1,4}$ are the volume fractions; λ_i, $i = \overline{1,5}$ are the thermal conductivity coefficients; μ is the dynamic viscosity coefficient; σ is the Stefan–Boltzmann constant; ω is the linear rate of coal-plastic thermochemical destruction. Subscripts and superscripts: e is the outer edge of the boundary layer; w is the surface of the body in flow; $*$ is the characteristic value; 0 are the initial conditions; s is the condensed phase; g is the gas phase; ∞ is the parameters at infinity; c is the carbon surface; 1–5 relate to the binder, pyrosol, coke, filler, and the gas, respectively; p is the pore; ef is the effective value.

Method of Calculation, Tests, and Initial Data. The system of Eqs. (1.1.7)–(1.1.12) with the initial and boundary conditions (1.1.15)–(1.1.18) was solved numerically using the implicit, absolutely stable monotone difference scheme [45]. The results found in [13] were reproduced for verifying the calculation program [13]. In addition, the numerical method was tested for the reference case (combined pyrolysis conditions). With all initial parameters being equal, the calculation was made for different spatial steps $h_1 = 10^{-4}$ m, $h_2 = 2 \cdot h_1$, $h_3 = h_1/2$, $h_4 = h_1/4$. The temperature of the thermal protective material and gas density ρ_5 was recorded across the depth of the body at different time points. In all cases, the problem was solved with a variable time step chosen on the assumption that the prescribed accuracy was equal for all spatial steps. The difference in body temperature and gas density of the thermal protective material ($\varepsilon = \max \, [\varepsilon_T, \varepsilon_{\rho_5}]$) decreased: $\varepsilon_1 = 10.8\%$, $\varepsilon_2 = 5.1\%$, $\varepsilon_3 = 1.9\%$.

The thermokinetic constants k_{iw}, E_{iw} in (1.1.3) for reactions (1.1.2) and (1.1.1) are presented in [9] and [44, 46], respectively. The enthalpy of graphite h_c was calculated by the formulas from [58]. For coal-plastic, the thermophysical coefficients c_{p4}, λ_4 and density ρ_4 were taken from [14]. The coefficients b_1, b_2 in the formulas (1.1.13) and (1.1.19) can be found in [59], while λ_{5*}, μ_* in the Eq. (1.1.14) as a function of time are given in [60]. Permeability (z) of the ideal porous medium in the Darcy law from (1.1.13) can be determined by well-known Karmana–Kozeny's formula [9]. The results presented below were found at $T_0 = 293$ K, $(\alpha/c_p)_0 = 0.2$ kg/(m^2 s), $A = 0.02$ kg/(m^2 s), $v = 10$–70 s^{-1}, $T_* = 1500$ K, $\mu_* = 4.2 \times 10^{-5}$ kg/(m s), $\lambda_{5*} = 0.067$ W/(m K), $h_e = 1.449 \times 10^7$ J/kg, $T_e = 3600$ K, $\rho_e = 0.088$ kg/m^3, $c_{p1} = c_{p2} = 1700$ J/(kg K), $c_{p3} = 1020$ J/(kg K), $\rho_1 = 1200$ kg/m^3, $\rho_2 = 1100$ kg/m^3, $\rho_3 = 1300$ kg/m^3, $\lambda_1 = \lambda_2 = 0.21$ W/(m K), $\lambda_3 = 0.041$ W/(m K), $L_0 = 10^{-2}$ m, $E_1/R = 3462$ K; $E_2/R = 11,305$ K; $k_1 = 9.6 \times 10^4$ s^{-1}, $k_2 = 1.2$ 10^5 s^{-1}, $q_1 = 2 \times 10^5$ J/kg, $q_2 = 10^5$ J/kg, $z_* = 6 \times 10^{-11}$m^2, $R = 8.314$ J/ (mol K), m$_1 = 16$ kg/kmol, m$_2 = 32$ kg/kmol, m$_3 = 14$ kg/kmol, m$_4 = 28$ kg/kmol, m$_5 = 28$ kg/kmol, m$_6 = 44$ kg/kmol, $M_5 = 29$ kg/kmol, $\sigma = 5.67 \times 10^{-8}$ W/ (m^2 K^4), $\alpha_1 = 0.5$, $\alpha_2 = 0.5$, $\varepsilon = 0.9$, $\varphi_{1,0} = 0.3$, $\varphi_{2,0} = 0.01$, $\varphi_{3,0} = 0.01$, $\varphi_{4,0} = 0.6$, $b_1 = 965.5$, $b_2 = 0.0735$.

Discussion of Numerical Solution Results. Figure 1.1, *I, II, III* show the time dependences of the surface temperature (a), heat transfer coefficient from (1.1.6) (b), and ablation (c) from the surface of the thermal protective material for the short time $t \leq = 0.5$ s. Figure 1.1, *I, II, III* correspond to frequencies $v = 10, 40,$ and 70 s^{-1}, respectively.

The detailed description of the surface temperature profile T_w shows that it is non-monotonic on some sections responding to fluctuations of the heat transfer coefficient (α/c_p) according to the formula (1.1.6).

The ablation rate $(\rho v)_w$ reaches its peak at the initial time ($t < 0.05$ s) because of the steep rise in T_w. Then, due to the generation of the coked layer and the increase in its resistance to the moving filtering gas, a value of $(\rho v)_w$ may decrease. However, as the thermal protective material is increasingly heated by the convective heat flow q_w and the heat wave penetrates deep into the body, the pyrolysis of coal-plastic continues, with gaseous and condensed products generated by the binder (resin) decomposition reaction. Since the maximum pressure of gaseous filtration products is reached inside the thermal protective material [13], these products can move deep into the body. Due to the exothermal reaction with coke generation from pyrosol (1.1.1), they can heat up coked and underlying cold layers of the thermal protective material. This leads to the further increase in $(\rho v)_w$. In addition, at $T_w > 800$ K, the ablation from the surface of the thermal protective material is continuous by virtue of the carbon surface destruction at kinetic $T_w < 1600$ K and diffusion $T_w > 1600$ K [4] conditions of heterogeneous chemical reactions (1.1.2).

It should be noted that at the pulsation rate of $10 \leq v \leq 70$ s^{-1}, the non-monotonic surface temperature profile arises (see Fig. 1.1, *I, II, III*) at short times $0 < t \leq 0.5$ s, responding to the sawtooth pattern of the heat transfer

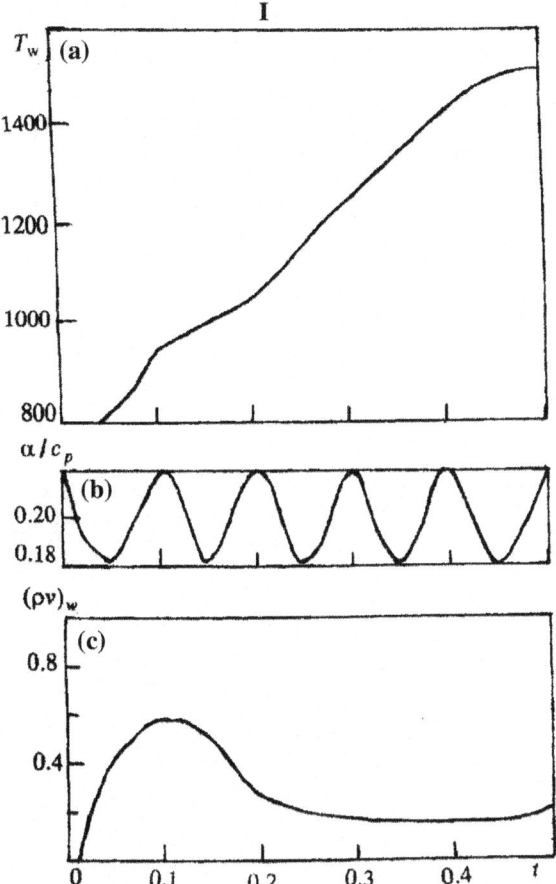

Fig. 1.1 Surface temperature (a), heat transfer coefficient (b), and ablation rate (c) as a function of time, for frequencies $\nu = 10 \ s^{-1}$ (*I*), $40 \ s^{-1}$ (*II*), $70 \ s^{-1}$ (*III*). *T*, K; (α/c_p), kg/(s m^2); $(\rho v)_w$, kg/(s m^2); *t*, s

coefficient. With the increase in the frequency $\nu = 10, 40, 70 \ s^{-1}$, the intensity of heat transfer with coal-plastic falls, while surface temperatures at $t = 0.5$ s are equal to $T_w = 1231, 1220, 1208$ K, respectively.

The latter result is in qualitative agreement with the experimental data [46]. The value T_w for the amplitude $A \neq 0$ and destruction time $0 < t \leq 10$ s is 10–15 K less than that at $A = 0$. The pulsating heat flow leads to more intensive thermal destruction of the binder and the increased injection of $(\rho v)_{1w}$. In this case, the surface temperature decreases due to reduction of the convective heat flow q_w in the formula (1.1.6) and, therefore, the total heat flow to the condensed phase.

$$Q_w = q_w - (\rho v)_{1w}(h_w - h_g) - (\rho v)_{2w}(h_w - h_c) - \varepsilon \sigma T_w^4. \qquad (1.1.20)$$

Figure 1.2a shows, at $\nu = 40 \ s^{-1}$, the time dependences of the surface temperature T_w, convective q_w, and total heat flows into the condensed phase Q_w (solid, dashed, and dot-and-dash lines, respectively). *1–3* correspond to heating of the solid

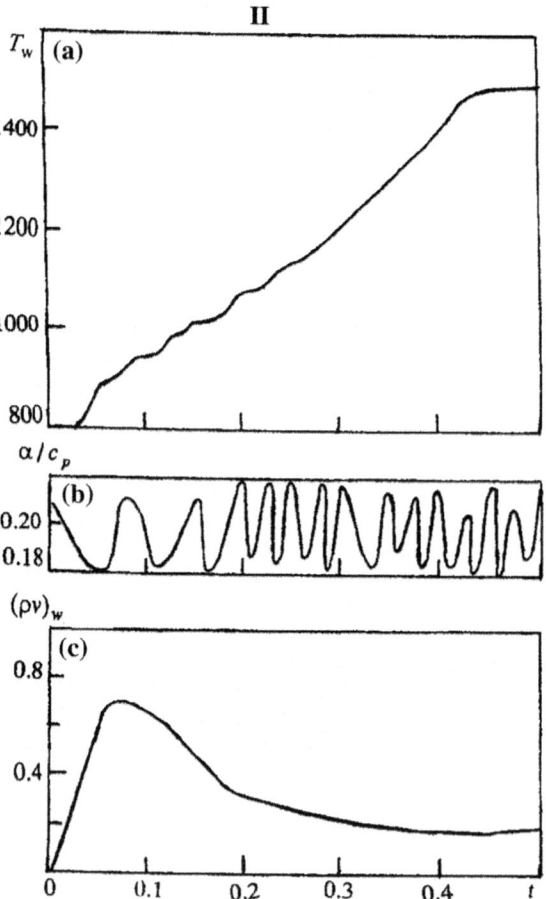

Fig. 1.1 (continued)

(inert) body, the two-stage (combined) pyrolysis process, and the one-stage decomposition reaction from (1.1.1).

Figure 1.2b demonstrates the time dependence of ablation rate. The experimental lines in Fig. 1.2a, b are marked with *4*. As shown in Fig. 1.2, the surface temperature at $t > 1$ in the second (combined) case is significantly lower as compared to *1* and *3*. This is associated with the large ablation in case *2* as compared to $(\rho v)_w$ in case *3* (see Fig. 1.2b) due to filtration of gaseous decomposition products. In addition, case *2* demonstrates the significant attenuation of the heat flow from the gas phase according to the formula (1.1.6) in contrast to case *3*, while in case *1* no attenuation is observed. In the third case, the enthalpy of the carbon wall h_c significantly increases when the surface of the thermal protective material is diffusively destroyed at $t > 4$ s. This enthalpy exceeds h_w in the third addend of the right-hand side of the formula (1.1.20) for the total heat flow (Fig. 1.2a—dot-and-dash lines *2* and *3*).

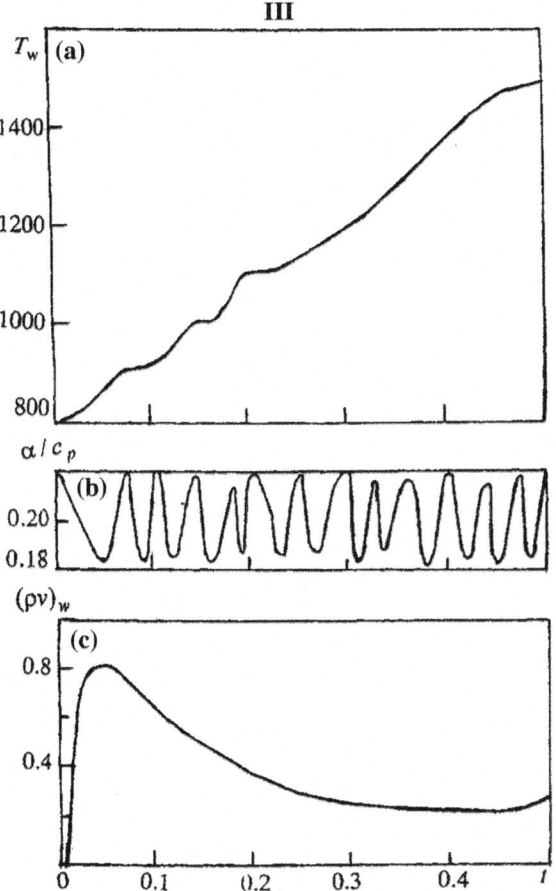

Fig. 1.1   (continued)

It should be noted that in the second case, the surface temperature T_w and the maximum ablation rate $(\rho v)_w$ are well consistent (up to an order of magnitude) with the known experimental results [46] for frequency $\nu = 40 \text{ s}^{-1}$. The difference between the calculated and experimental data for the surface temperature does not exceed 16–18%.

Figure 1.3a presents the temperature distribution of coal-plastic, while Fig. 1.3b shows the volume fraction of the coke as a function of the layer depth at different time points. The solid lines in Fig. 1.3a correspond to the combined coal-plastic destruction. The dashed lines stand for the inert heating of the solid body, with other initial parameters being equal. The effectiveness of pyrolysis as a heat barrier is associated with slower coal-plastic thermal destruction due to a thicker pro-coked layer (see Fig. 1.3b) and a lower temperature inside the body. This is consistent with the conclusion in [14].

Fig. 1.2 Surface temperature
(**a**) and ablation rate (**b**) as a
function of time for
$v = 40$ s^{-1}: *1–3* correspond to
heating of the solid (inert)
body, the two-stage
(combined) pyrolysis process,
and the one-stage
decomposition reaction from
(1.1.1), respectively: *4*—to
the experiment

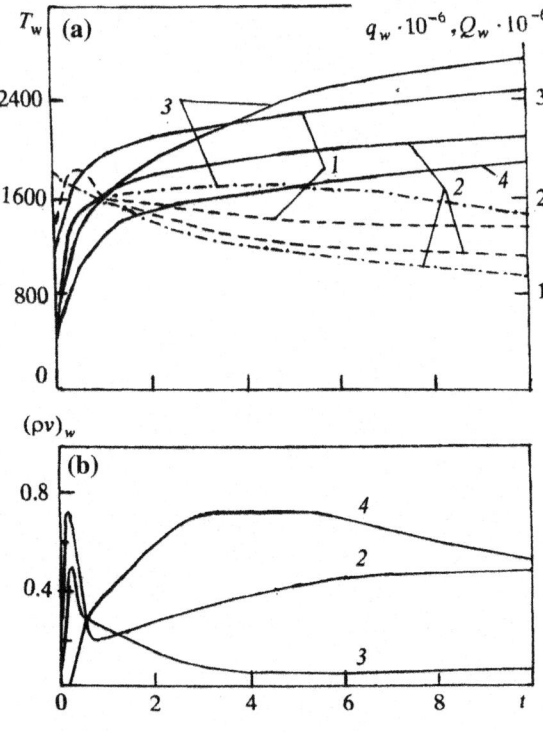

Fig. 1.3 coal-plastic
temperature distribution (**a**),
coke fraction (**b**) over the
depth of the layer at times: *1*
—1 s; *2*—5 s; *3*—10 s. The
solid lines in **a** correspond to
the combined destruction of
coal-plastic, dashed lines—to
inert heating of the solid body

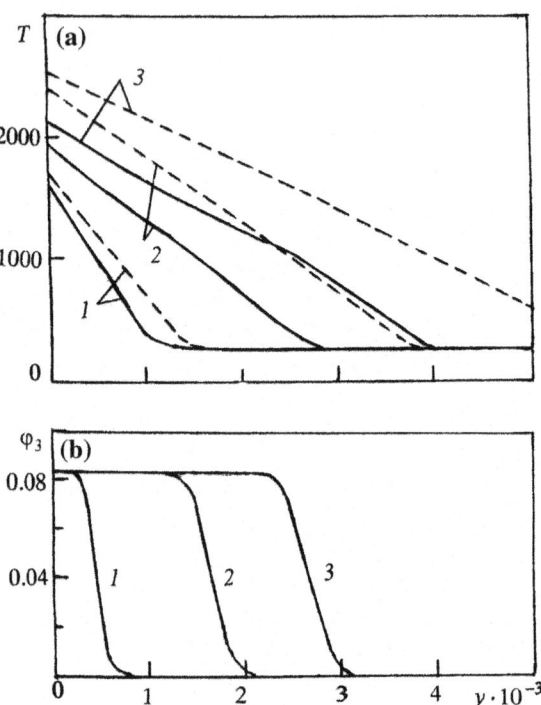

1.2   Mathematical Modeling of Heat and Mass Transfer in Thermal Protective Coatings with Gas Flow Pulsation

This section studies the effects of periodic perturbations and wall vibrations on the intensity of coal-plastic thermochemical destruction as well as compares the obtained results with the known data.

The physical nature of oscillations is the wall vibration toward an incident flow. Wall vibrations were generated according to the harmonic law using a special vibration exciter [44, 52].

The comparison of filtration and thermal characteristics of porous materials analyzed in the presence of pulsating and vibrational perturbations [52] demonstrates that dependences of the viscous term in the filtration law and the relative heat transfer function on oscillation intensity and hydrodynamic nature of the heat transfer process are relatively similar to each other. The expression for additional heat transfer q' in a porous body with periodic pulsations of a gas coolant was obtained in [51, 52]:

$$q' = -\frac{c_{p5}\rho_5\varphi_5 B^2 f}{2\sqrt{2}}\frac{\partial T}{\partial y},\qquad(1.2.1)$$

where c_{p5}, ρ_5 are, respectively, the specific heat capacity and the true density of the gas phase of the thermal protective material; $f = 2\pi\nu$; B, ν are the amplitude of wall pulsations and the frequency of periodic perturbations. The prime symbol stands for gas flow pulsations. λ_4 in (1.1.13) can be found by the formula [2, 44].

$$\lambda_4 = \lambda_{4c} + \lambda_4', \lambda_4' = \frac{\pi\nu c_{p5}\rho_5\varphi_5 B^2}{\sqrt{2}}.\qquad(1.2.2)$$

The expression for effective viscosity μ_{ef} in the Darcy law (1.1.13) can be taken as the Einstein's modification [61]:

$$\mu_{ef} = \mu[1 + C\cos(tf)],\qquad(1.2.3)$$

where C is the dimensionless coefficient; μ is the viscosity of gaseous filtration products without periodic disturbances.

For the pulsating flow case, an increase or decrease in viscosity is associated with additional energy dissipation due to redistribution of temperature, pressure, and other gradients.

The calculation results presented below were obtained by the numerical solution of the boundary problem (1.1.7)–(1.1.12), (1.1.15)–(1.1.18), where μ_{ef} and the thermal conductivity coefficient in (1.1.13), (1.1.14) were taken from the Eqs. (1.2.2 and 1.2.3), while the specific heat capacity (c_{p4}) from [14]. These results were obtained at $\nu = 10$–30 s^{-1}, $B = 2 \times 10^{-3}$ m, $0 < C \le 0.25$.

Discussion of Numerical Solution Results. Figure 1.4 shows the surface temperature T_w (a) and the total rate of ablation $(\rho v)_w$ (b) from the surface of the thermal protective material as a function of time. Lines *1–3* in Fig. 1.4 correspond to frequencies $v = 10, 20$, and 30 s^{-1}, respectively. The ablation rate $(\rho v)_w$ reaches its peak at a time close to the initial one ($t < 0.05$ s) because of the steep rise in T_w. Then, due to the generation of the coked layer and the increase in its resistance to the moving filtering gas, a value of $(\rho v)_w$ may decrease. However, as the thermal protective material is increasingly heated by the convective heat flow q_w and the heat wave penetrates deep into the body, the pyrolysis of coal-plastic continues, with gaseous and condensed products generated by the binder (resin) decomposition reaction. Since the maximum pressure of gaseous filtration products is reached inside the thermal protective material [13], these products can move deep into the body. Due to the exothermal reaction with coke generation from the pyrosol (1.1.1), they can heat up coked and underlying cold layers of the thermal protective material. This leads to the further increase in $(\rho v)_w$. In addition, at $T_w > 800$ K, ablation from the surface of the thermal protective material is continuous by virtue of the carbon surface destruction at kinetic $T_w < 1600$ K and diffusion $T_w > 1600$ K [4] conditions of heterogeneous chemical reactions (1.1.2).

With the increase in frequency $v = 10, 20$, and 30 s^{-1}, the intensity of heat transfer with coal-plastic falls (see Fig. 1.4a), while the intensity of mass transfer increases (see Fig. 1.4b). The latter result is in qualitative agreement with the

Fig. 1.4 Temperature dependence of the surface (**a**) and total mass velocity (**b**) on time. Curves *1–3* responsible largest frequency $v = 10, 20, 30$ s^{-1}

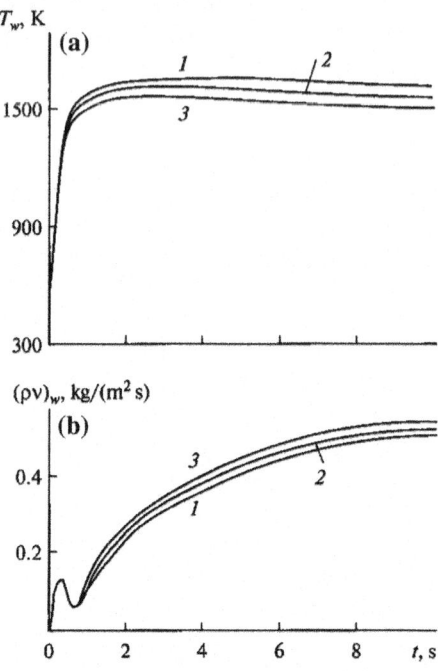

experimental data [46]. The pulsating heat flow leads to more intensive thermal destruction of the binder and increased injection of $(\rho v)_{1w}$.

In this case, the surface temperature decreases due to reduction of the convective heat flow q_w in the formula (1.1.6) and the total heat flow to the condensed phase:

$$Q_w = q_w - (\rho v)_{1w}(h_w - h_g) - (\rho v)_{2w}(h_w - h_c) - \varphi_4 \varepsilon \sigma T_w^4.$$

On the steady-state section of the coal-plastic thermochemical destruction ($8 \leq t \leq 10$ s), the difference between the calculated and experimental surface temperatures (see Fig. 1 in [46]) at $v = 10$ s^{-1} and $v = 30$ s^{-1} does not exceed 22 and 30%, respectively.

In case of unsteady coal-plastic heating ($t < 8$ s), the physical experiment showed transient and turbulent filtrations with temperature increase [46]. In addition, the non-monotonic change in $(\rho v)_w$ was observed in the course of time [46], with the behavior and the value depending on decomposition rate of the coal-plastic binder and filler. At the first stage of the study, these physical processes (turbulent flow in pores, peeling and delamination of the filler) associated with heating of the thermal protective material were ignored in the mathematical model. This is explained by the lack of reliable data on filtration, structural, and kinetic parameters of the used materials [14, 46].

Figure 1.5 shows the distribution of coal-plastic temperature over the depth of the layer at different time points for $v = 10$ s^{-1}.

Fig. 1.5 Temperature distribution over the layer depth in various time points for $v = 10$ s^{-1}. The solid lines correspond to $\lambda_4' = 0$, $C = 0$; the dashed lines—to $\lambda_4' \neq 0$, $C = 0$; the dot-and-dash lines —to $\lambda_4' \neq 0$ $C \neq 0$, respectively. The lines marked with *1–3* are given for times $t = 1, 5, 10$ s

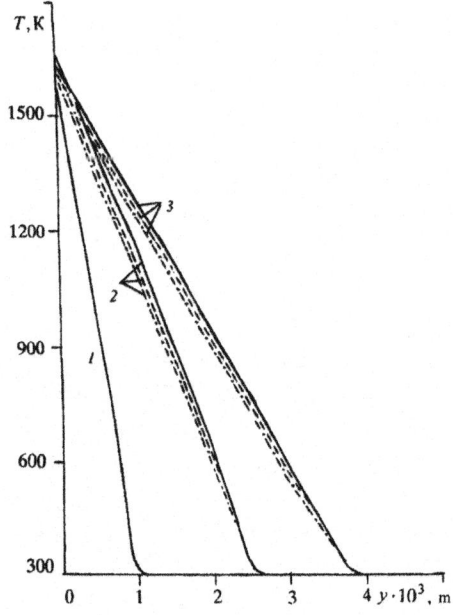

Table 1.1 Temperature of coal-plastic depending on parameters of the pulsating flow at $t = 10$ s

		T (K)	
y, m	0	2×10^{-3}	3×10^{-3}
$\lambda_4' = 0,\ C = 0$	1647	773	433
$\lambda_4' \neq 0,\ C = 0$	1615	758.3	406
$\lambda_4' \neq 0,\ C = 0.15$	1627	760	408
$\lambda_4' \neq 0,\ C = 0.2$	1614	758.6	413
$\lambda_4' \neq 0,\ C = 0.25$	1591	757.4	420

The solid lines in Fig. 1.5 correspond to destruction of the thermal protective material without pulsating components of heat transfer $\lambda_4' = 0$ in (1.2.3) and viscosity $C = 0$ in (1.2.2). The dashed lines relate to the case: $\lambda_4' \neq 0$, $C = 0$; the dot-and-dash lines—to the combined heat and mass transfer: $\lambda_4' \neq 0$, $C = 0.2$. The lines marked with *1–3* are given for time $t = 1, 5, 10$ s.

The analysis of the lines in Fig. 1.5 suggests that the wave coefficient of thermal conductivity λ_4' arising in the permeable material exposed to fluctuations reduces the temperature of the thermal protective material due to the increasing intensity of filtration flows.

Such an effect is further intensified in the combined heat and mass transfer. $\lambda_4' \neq 0$, $C \neq 0$ (see Fig. 1.5). This method of heat removal seems to be feasible in case of exothermic reactions in ablative materials.

Table 1.1 presents temperatures of coal-plastic on the surface and deep in the body at $t = 10$ s depending on pulsating components λ_4' and C in formulas (1.2.3) and (1.2.2).

It follows from Table 1.1 that being far from the surface, $y = 3 \times 10^{-3}$ m, the temperature of coal-plastic varies monotonically with increase in C, and the heat wave slightly warms the sample.

However, the temperature of the surface and near it, $y = 2 \times 10^{-3}$ m, may be non-monotonic: firstly increase and then decrease with growth of C. The latter result seems to be associated with additional dissipation of the energy caused by the pulsating flow.

1.3 Mathematical Modeling of Surface Roughness and Ablation Effects on Thermal Protection

For high-enthalpy flows, surface condition is one of the factors that affects heat flow (heat exchange coefficient) and determines the heat transfer rate [3, 4, 62, 63]. Ablating composite materials, such as glass and coal-plastic [2, 4] consisting of a filler and a binder, can have surface roughness in a wide range: $(0.05–0.5) \times 10^{-3}$ m. This roughness value is high enough to change a laminar boundary layer to a turbulent layer [3, 62, 63].

The development of aerospace technology has shown [3, 64–66] that roughness should be taken into account for ablative head sections of streamlined bodies and for ablative side surfaces with a relief surface structure. The relief surface structure is generated during ablation in a supersonic turbulent flow, which is confirmed in [66].

A rough surface may lead to a significant growth in surface friction and an increase in convective heat flow [64]. Roughness-induced growth of a heat flow in a turbulent boundary layer plays an important role in determining ablation of thermal protective material and changing the shape of an aircraft head [64–66].

Problem Statement. Since many experiments [3, 64–66] show that surface friction increases with growing roughness height and heat flow reaches its maximum value [64, 65], the Reynolds analogy seems not to apply to rough surfaces. We can use the findings of the study [65] which presents a new approach to engineering calculations of heat flows to rough surfaces in supersonic streams:

$$\frac{St}{St_*} = \frac{C_f/C_{f*}}{1 + \eta\, F}, \qquad \frac{C_f}{C_{f*}} = \lg(Re_K)\left(0.365\frac{h_w}{h_r} + 0.635\right), \qquad (1.3.1)$$

$$F = (Pr)^{0.8}(Re_K)^{0.45}\left(\frac{C_f}{C_{f*}}\right)^{0.725}\frac{U_{\tau*}}{U_e}, \qquad \frac{U_{\tau*}}{U_e} = \left(\frac{C_{f*}}{2}\frac{\rho_e}{\rho_w}\right)^{0.5},$$

$$Re_K = \frac{U_{\tau*}K}{\nu_*}, \qquad St_* = \frac{(\alpha/c_p)_*}{(\rho\, U)_e}$$

$$St = \frac{q_w}{(\rho\, U)_e(h_e - h_w)}, \qquad Pr = \left(\frac{c_p\mu}{\lambda}\right)_e, \qquad (1.3.2)$$

$$U_\tau = \left(\frac{\tau_w}{\rho_w}\right)^{0.5}, \qquad \nu = \frac{\mu}{\rho},$$

$$\chi = \frac{c_p}{c_v}, \qquad c_{pg} = b_1 + 2b_2 T, \qquad h_r = c_{pg}T_r,$$

$$T_r = T_\infty[1 + 0.5r(\chi - 1)M_\infty^2], \qquad C_{f*} = 0.0162(K/l)^{1/7}$$

where Pr is the Prandtl number; M_∞ is the Mach number; St is the Stanton number; Re_K is the Reynolds number for roughness; C_f is the local friction coefficient; l is body length; h is enthalpy; U is the velocity; r is the coefficient of restitution; τ is shear stress; χ is the heat capacity ratio; η, b_1, b_2 are constants; K is the equivalent roughness height [63, 65]. Index τ is the dynamic velocity.

The exponential resistance law [the last formula in Eq. (1.3.2)] derived for steady-state roughness [67] was the total resistance coefficient, depends only on local relative roughness, and does not depend on the Reynolds number.

Let us define the heat flow $q_w(t)$ acting on the thermal protective material for a definite period of time. Then, according to [4, 65, 67] from (1.3.1), (1.3.2), we obtain:

$$q_w = \frac{\alpha}{c_p} \times \frac{C_f/C_{f*}}{1+\eta\,F}\,(h_e - h_w),$$

$$\frac{\alpha}{c_p} = \left(\frac{\alpha}{c_p}\right)_0 \left[1 - \frac{\gamma(\rho v)_w}{(\alpha/c_p)_0}\right]$$

(1.3.3)

where $(\rho v)_w$ is the total ablation of the thermal protective material and γ is the attenuation factor for a turbulent flow in a boundary layer.

The following results were obtained in the numerical solution of the problem (1.1.7)–(1.1.12), (1.1.15)–(1.1.18) for the input data of the first section. Heat flow q_w in the formula (1.1.15) was from (1.3.3); the findings presented below were obtained at $K = (0.127–0.508) \times 10^{-3}$ m, $t_p = 10$ s, $Pr = 0.9$, $Re_K = 15–65$, $M_\infty = 5$, $\chi = 1.4$, $r = 0.9$, $\eta = 0.52$.

Discussion of the Numerical Solution Results. At first, flowing to coal-plastic is considered without pyrolysis, taking into account only heterogeneous reactions (1.1.2): $(\rho v)_w = (\rho v)_{2w}$. Figure 1.6 presents the time dependencies of the surface temperature T_w. The solid lines *1–4* stand for the equivalent roughness height: $K = 1.27 \times 10^{-4}, 2.547 \times 10^{-4}, 3.81 \times 10^{-4}, 5.08 \times 10^{-4}$ m, the dashed line for $K = 0$ (smooth surface). Figure 1.7 presents the distribution of the friction coefficient (line *1*) and the Stanton number (line *2*) depending on the equivalent roughness height at $t = t_p$.

As follows from Figs. 1.6 and 1.7, with an increase in equivalent roughness height (friction coefficient), the Stanton number (1.3.1) and, therefore, the convective heat flow from (1.3.3) and thermal protective material surface temperature also increase. This result is qualitatively consistent with the experimental data [63, 65].

Fig. 1.6 Time dependence of surface temperature. The solid lines represent the equivalent roughness height K: *1*— 1.27×10^{-4}, *2*— 2.54×10^{-4}, *3*— 3.81×10^{-4}, *4*— 5.08×10^{-4} m; the dashed line—$K = 0$ at $(\rho v)_w = (\rho v)_{2w}$

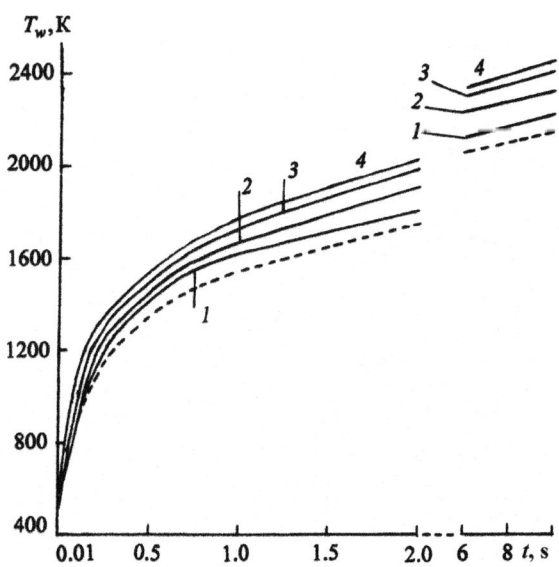

Fig. 1.7 Distribution of coefficient of friction (line *1*) and the Stanton number (line *2*) depending on equivalent roughness height at $t = t_p$

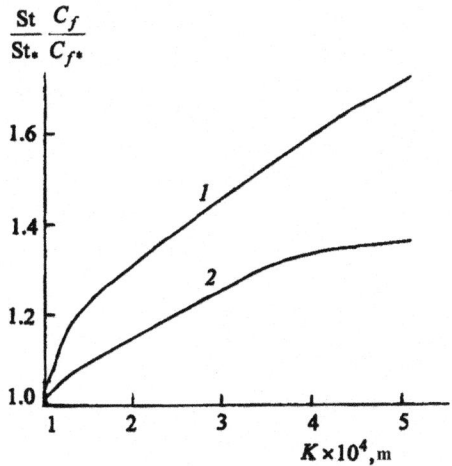

Let us consider heating the thermal protective material (see Fig. 1.8) in combined destruction mode—coal-plastic—taking into account the surface reactions (Fig. 1.9): $(\rho v)_w = (\rho v)_{1w} + (\rho v)_{2w}$, where the designations of the curves in Fig. 1.8 coincide with the designations in Fig. 1.6. As expected, with an increase in the equivalent roughness height, the surface temperature of the thermal protective material also increases.

This is qualitatively confirmed by the experiments [68, 69]. It was found in [68] that the roughness effect at $K < 10^{-4}$ m is insignificant—while the heat exchange is almost equivalent to the heat exchange for a smooth surface.

Fig. 1.8 Time dependence of surface temperature in case of complete ablation $(\rho v)_w = (\rho v)_{1w} + (\rho v)_{2w}$. The symbols are the same as in Fig. 1.6

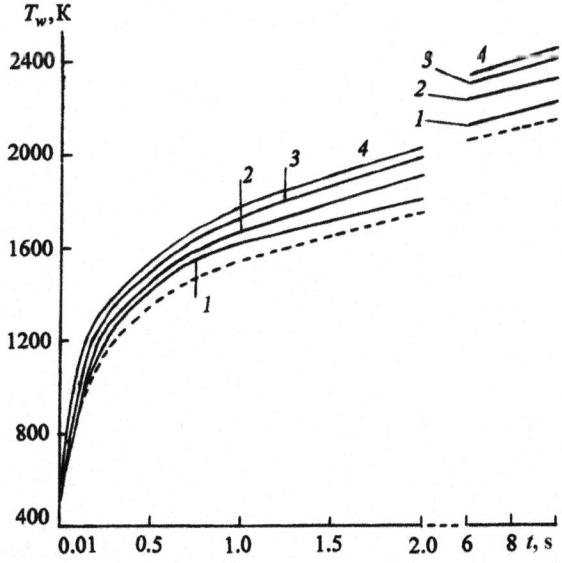

Under the effect of a high-temperature gas flow at $K > 5 \times 10^{-4}$ m, separate elements with marked roughness break off (chemical-mechanical chipping) and are taken into the gas flow. Since coal-plastic components are not equally subject to ablation, the boundary layer can be locally disrupted, and, as a result, thermal protection efficiency may be affected.

Comparison of the models in combined destruction mode with and without roughness is of interest. Figure 1.9 shows the time distribution of surface temperature (b) and ablation (a) where $K = 0$ (curve *1*) and $K = 1.27 \times 10^{-4}$ m (lines *2*). A decrease in surface temperature at $K \neq 0$, which takes place at the initial coal-plastic heating stage, is explained by a large injection (see Fig. 1.9a curve *2*) of gaseous thermochemical destruction products.

The experiment [68] shows that the surface temperature decreases in the combined destruction mode when coal-plastic is heated. This result in [68] was attributed to an increased contact surface due to roughness and, therefore, to an increased yield of volatile gaseous products generated during pyrolysis and chemical reactions of volatile components of a binder and filler.

The efficiency of pyrolysis as a thermal barrier at the initial stage of material destruction becomes apparent by the fact that the convective heat flow from the gas phase is significantly blocked. Then, as the heat exchange process reaches the steady state at $t > 0.5$ s, hydrodynamic resistance increases. As the thickness of the coked coal-plastic layer grows [13, 14] and the pressure gradient of the gas in pores falls [13], the ablation value is stabilized. As a result, the Stanton number (1.3.1) begins to increase at $t > 3$ (see line *3* in Fig. 1.9a), along with a convective heat flow as defined by formula (1.3.3).

Fig. 1.9 Time distribution of surface temperature (**a**) and ablation (**b**) for of complete ablation. Lines *1* stand for $K = 0$, *2*—$K = 1.27 \times 10^{-4}$ m

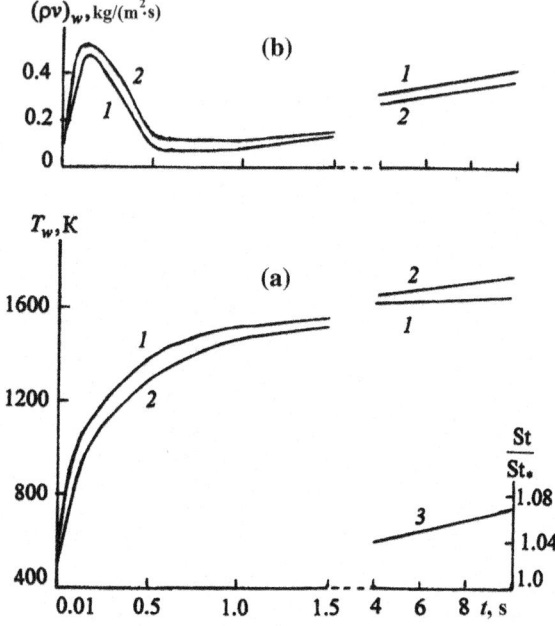

This leads to increased heating and growing coal-plastic surface temperature in the presence of roughness as compared with a smooth surface.

1.4   Thermal Protection of Multilayer Structure Under Fire Exposure

The commercial production of decorative structures imposes some additional technical requirements on construction materials. Lungs porous metals, wood, and polymer materials, such as fiberglass and coal-plastic, are best suited for these requirements.

However, these materials are sensitive to high temperatures and fire. For example, some alloy-steel structures are deformed and lose their stability and bearing ability after 15 min of intensive fire or test fire [70]. Some porous metals are rapidly heated and lose their strength properties at 500–700 K. Wood and plastics burn at 470 K and produce a significant amount of smoke and toxic substances [70].

Intumescent fire-retardant coatings are used for protection against fire in many industries: construction, cars, ships, aircrafts. In Russia, the most popular intumescent fire-retardant coatings are SKG-1, VPM-2, OVR-1, and 336–11–88 [71–73]. Their key advantage is the ability to greatly (by 20–40 min) increase thickness of a protective layer when they are heated and form a porous structure (coked cellular material) of low thermal conductivity.

There are a few studies (see [74] for overview) in the Russian and available foreign literature which are devoted to modeling heat and mass transfer in such systems [75–79]. The model presented in [75, 79] takes into account a variety of processes: heat emission and absorption during pyrolysis and evaporation, volumetric change, destruction of materials during swelling. That is why the mathematic model only for the intumescent fire-retardant coatings layer includes many nonlinear equations in partial derivatives. A simpler mathematical model based on the experimental findings in [76] was developed in [78] and tested in [80] for calculating thermal protection of multilayer containers under fire conditions.

The model proved itself to be efficient for calculating of how a structure is heated in laboratory and bench tests [78, 80].

Based on the models [4, 57, 80, 81], this paper examines the thermal regime of porous steel, being a part of a multilayer structure protected against fire by intumescent fire-retardant coatings (SKG-1 [72]).

Problem Statement. The arrangement of layers in the structure under study (tablet inside edging) in the cylindrical coordinate system (z, r) is shown in Fig. 1.10. The first heat-insulating layer (substrate) is made of a poorly conducting material (asbestos cement). Porous steel is used as the second composite material. Finally, the third layer (facing) is the intumescent coating SGK-1.

Fig. 1.10 Heat transfer of the body with the external environment

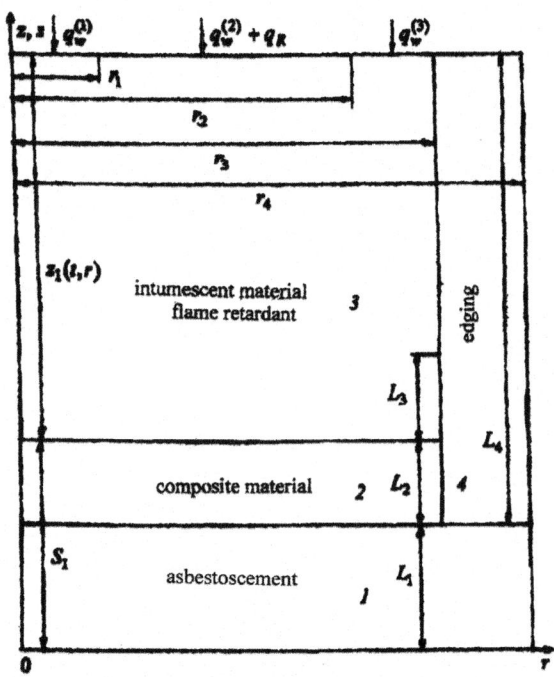

Thermal processes in the distortion-free second layer—porous metal—and the deformable third fire-retardant layer are considered within the framework of the conjugate mathematical model for non-stationary heat and mass transfer [80]. Intumescent fire-retardant coatings are simulated by a permeable multiphase single-temperature (due to low temperatures $T < 1200$ K) reacting environment which contains condensed and gas phases. The material loses its mass during heating, and its expansion is assumed to be irreversible and proceeds in a one-dimensional way along the coordinate z. The studies [78, 80] show that such a destruction of intumescent fire-retardant coatings is consistent with the real process with a good accuracy.

However, in contrast to [78, 80], this paper proposes to take heat transfer by conduction along the transverse coordinate r [77]. Heat by conduction along r is taken into account for modeling heat and mass transfer both in porous steel [4, 57] and substrate.

The mathematical problem defined above on the assumptions for the third layer (intumescent fire-retardant coatings) in the Lagrange coordinate system s, r, t (intumescence takes place only along the axis z) is expressed as [77, 78, 80].

$$\rho_0 \frac{\partial \bar{m}}{\partial t} + \frac{\partial G_g}{\partial s} = 0, \quad s\epsilon(S_1, z_1(t, r)), \quad S_1 = L_1 + L_2, \quad r\epsilon(0, r_3), \quad 0 < t \le t_k \quad (1.4.1)$$

$$(\rho c_p)_{ef} \frac{\partial T_3}{\partial t} + \frac{(Gc_p)_g}{f} \frac{\partial T_3}{\partial s} = \frac{1}{f} \frac{\partial}{\partial s}\left(\frac{\lambda_{3,\parallel}}{f} \frac{\partial T_3}{\partial s}\right) + \frac{1}{r} \frac{\partial}{\partial r}\left(r\lambda_{3,\perp} \frac{\partial T_3}{\partial r}\right)$$
$$+ \frac{\rho_0 Q}{f} \frac{\partial \bar{m}}{\partial t} \qquad (1.4.2)$$

$$s\in(S_1, z_1(t,r)), \quad r\in(0, r_3), 0<t\le t_k,$$
$$\frac{\partial \bar{m}}{\partial t} = -k(\bar{m} - \bar{m}_*)\exp\left(-\frac{E}{RT_3}\right), \quad 0<t\le t_k \qquad (1.4.3)$$

$$(\rho c_p)_{ef} = (\rho\, c_p\varphi)_T + (\rho\, c_p\varphi)_g, \quad c_{pg} = b_1 + 2b_2 T_3, \quad \Delta m = m_0 - m,$$

$$z(t,r) = \int_{S_1}^{s} f(t,r,y)dy, \quad f = 1 + \theta, \theta = \frac{\Delta z}{L_3}, \quad \Delta z = z_1 - L_3$$

$$\bar{m} = \frac{\Delta m}{m_0}, \varphi_T = \frac{\rho_0 \bar{m}}{\rho_T f}, \quad \varphi_g = 1 - \varphi_T, \quad \lambda_{3,\parallel} = (\lambda\varphi)_T + \varphi_g(\lambda_R + \lambda_g),$$

$$\lambda_R = 2\varepsilon_R^2 \sigma\, d_0 f T^3, \quad \lambda_{2,\perp}^{-1} = \frac{\varphi_T}{\lambda_T} + \frac{\varphi_g}{\lambda_g + \lambda_R} \qquad (1.4.4)$$

$$z_1(t,r) = \int_{S_1}^{S_2} f(t,r,y)dy, \quad S_2 = L_1 + L_2 + L_3 \qquad (1.4.5)$$

where z_1 is the coordinate of intumescent fire-retardant coatings external surface; ρ_T, ρ_0 is the density of a building frame material and of a material at the initial instant; φ_T, ρ_g are their volume fractions; θ is the degree of intumescent fire-retardant coatings expansion; x is the current size of intumescent fire-retardant coatings; Δz is a change in intumescent fire-retardant coatings size; c_p is the specific heat capacity; λ is the thermal conductivity coefficient; T is the temperature; G is the mass flow rate of gaseous products generated from intumescent fire-retardant coatings thermal destruction; t is the time; Δm is a change in intumescent fire-retardant coatings mass; m is a relative change in intumescent fire-retardant coatings mass during heating; E, k, Q are, respectively, the energy of activation, the preexponential factor, and the thermal effect of intumescent fire-retardant coatings thermal destruction reaction; ε_R is the emissivity factor; σ is the Stefan–Boltzmann constant; R is the universal gas constant; d_0 is the initial diameter of intumescent fire-retardant coatings pores; L_i, $i = 1, 2, 3$ is the thickness of layers of a multilayer material along the axis z; r_3 is the radius of the three-layer tablet as shown in Fig. 1.10; b_1, b_2 are constants. Subscripts and superscripts: g is the gas phase; T is the condensed phase; 1, 2, 3, 4 subscripts are, respectively, temperatures of asbestos cement, composite material, intumescent fire-retardant coatings, and edging; R is the radiant component of thermal conductivity coefficient; ef are effective parameters; 0 is the initial value; $*$ is the characteristic quantity; $\parallel$ is parallel to the flow direction; $\perp$ is perpendicular to the flow direction; top feature is dimensionless quantities; m is metal.

It is assumed that in the permeable metal, the gas phase motion over the length and breadth of the second layer is absent: $v|_s = 0$ (v is gas filtration rate; S is second boundary layer). This is due to the absence of pressure gradient at $z = L_1$, $r = 0$, $r = r_3$ ($v_{L_1} = v_0 = v_{r_3} = 0$) and a small change in pressure: $P_{S_1} \approx P_0$ due to low temperatures $T_2 < 350$ K (see Fig. 1.12b). Then for the second porous layer (metal) compound tablets, heat equation has the form [2, 4]

$$c_{p2}\frac{\partial T_2}{\partial t} = \frac{\partial}{\partial z}\left(\lambda_{2,\|}\frac{\partial T_2}{\partial z}\right) + \frac{1}{r}\frac{\partial}{\partial r}\left(r\lambda_{2,\perp}\frac{\partial T_2}{\partial r}\right),$$

$$L_1 < z \le S_1, \quad r\epsilon(0, r_3), \quad 0 < t \le t_k,$$

(1.4.6)

$$c_{p2} = \rho_{2m}c_{pm}(1 - \varphi) + \rho_{2g}c_{pg}\varphi,$$

$$\lambda_{2,\|} = \lambda_{2m}(1 - \varphi) + \lambda_{2g}\varphi, \quad \lambda_{2,\perp}^{-1} = \frac{1 - \varphi}{\lambda_{2m}} + \frac{\varphi}{\lambda_{2g}}$$

For example, the thermal conductivity equations in an insulant (asbestos cement) and edging can be written as:

$$(\rho c)_{ac}\frac{\partial T_1}{\partial t} = \lambda_{ac}\frac{\partial^2 T_1}{\partial z^2} + \frac{\lambda_{ac}}{r}\frac{\partial}{\partial r}\left(r\frac{\partial T_1}{\partial r}\right),$$

$$z\epsilon(0, L_1), \quad r\epsilon(0, r_4), \quad 0 < t \le t_k$$

(1.4.7)

$$(\rho c)_{ok}\frac{\partial T_4}{\partial t} = \lambda_{ok}\frac{\partial^2 T_4}{\partial z^2} + \frac{\lambda_{ok}}{r}\frac{\partial}{\partial r}\left(r\frac{\partial T_4}{\partial r}\right),$$

$$z\epsilon(L_1, L_1 + L_4), \quad r\epsilon(r_3, r_4), \quad 0 < t \le t_k$$

(1.4.8)

The system of Eqs. (1.4.1)–(1.4.3), (1.4.6)–(1.4.8) must be solved taking into account the following initial and boundary conditions.

Initial conditions

$$T_i|_{t=0} = T_0, \quad i = 1, 2, 3, 4, \quad \bar{m}|_{t=0} = 1$$

(1.4.9)

At $z = z_1(t, r)$, the external surface of the tablet produces non-uniform heating from the gas phase, where the heat convective flow $q_w^{(i)} = (\alpha/c_p)^{(i)} \cdot (h^{(i)} - h_w)$ $i = 1, 2, 3$ is defined as:

$$\left(\frac{\alpha}{c_p}\right)^{(1)}(h^{(1)} - h_w) = -\frac{\lambda_{3,\|}}{f}\frac{\partial T_3(z_1, r, t)}{\partial s}, \quad 0 \le r < r_1,$$

(1.4.10)

$$\left(\frac{\alpha}{c_p}\right)^{(2)}(h^{(2)} - h_w) + q_R = -\frac{\lambda_{3,\|}}{f}\frac{\partial T_3(z_1, r, t)}{\partial s}, \quad r_1 \le r \le r_2,$$

(1.4.11)

$$\left(\frac{\alpha}{c_p}\right)^{(3)} (h^{(3)} - h_w) = -\frac{\lambda_{3,\|}}{f} \frac{\partial T_3(z_1, r, t)}{\partial s}, \quad r_2 < r \leq r_3 \tag{1.4.12}$$

$$h_w = c_{pg} T_{3w}, \quad \left(\frac{\alpha}{c_p}\right)^{(i)} = \left(\frac{\alpha}{c_p}\right)^{(i)}_{no} \left[1 - \frac{\gamma G_w}{(\alpha/c_p)^{(i)}_{no}}\right], \quad i = 1, 2, 3,$$

$$q_R = \varepsilon\sigma(T_e^4 - T_{3w}^4), \quad \varepsilon = \frac{1}{\varepsilon_e^{-1} + \varepsilon_w^{-1} - 1} \tag{1.4.13}$$

The conditions for mass conservation of the gas phase at $z = S_1$, ideal contact for temperatures at $z = S_1$, $z = L_1$, $r = r_3$, as well as impermeability of the gas phase at $z = L_1$:

$$G|_{z=S_1+0} = (\rho_5 v \varphi_5)|_{z=S_1-0}, \quad 0 \leq r \leq r_3, \tag{1.4.14}$$

$$\lambda_{2,\|} \frac{\partial T_2}{\partial z}\Big|_{z=S_1-0} = \frac{\lambda_{3,\|}}{f} \frac{\partial T_3}{\partial s}\Big|_{s=S_1+0},$$

$$T_2|_{z=S_1-0} = T_3|_{s=S_1+0}, \quad 0 \leq r \leq r_3 \tag{1.4.15}$$

$$\lambda_{ac} \frac{\partial T_1}{\partial z}\Big|_{z=L_1-0} = \lambda_{2,\|} \frac{\partial T_2}{\partial z}\Big|_{s=L_1+0},$$

$$T_1|_{z=L_1-0} = T_2|_{z=L_1+0}, \quad 0 \leq r \leq r_3 \tag{1.4.16}$$

$$\lambda_{ac} \frac{\partial T_1}{\partial z}\Big|_{z=L_1-0} = \lambda_{ok} \frac{\partial T_4}{\partial z}\Big|_{s=L_1+0},$$

$$T_1|_{z=L_1-0} = T_4|_{z=L_1+0}, \quad r_3 < r \leq r_4 \tag{1.4.17}$$

$$\lambda_{2,\perp} \frac{\partial T_2}{\partial r}\Big|_{r=r_3-0} = \lambda_{ok} \frac{\partial T_4}{\partial r}\Big|_{r=r_3+0},$$

$$T_2|_{r=r_3-0} = T_4|_{r=r_3+0}, \quad L_1 < z \leq S_1 \tag{1.4.18}$$

Across the cylinder, we set the symmetry condition at $r = 0$:

$$\frac{\partial T_i}{\partial r}\Big|_{r=0} = 0, \quad i = 1, 2, \ 0 \leq z \leq S_1,$$

$$\frac{\partial T_3}{\partial r}\Big|_{r=0} = 0, \quad S_1 < z \leq z_1 \tag{1.4.19}$$

and the heat transfer condition according to Newton's law at $r = r_4$:

$$-\lambda_{ac} \frac{\partial T_1}{\partial r}\Big|_{r=r_4} = \delta(T_1|_{r=r_4} - T_0), \quad 0 \leq z \leq L_1,$$

$$-\lambda_{ok} \frac{\partial T_4}{\partial r}\Big|_{r=r_4} = \delta(T_4|_{r=r_4} - T_0), \quad L_1 < z \leq L_4 \tag{1.4.20}$$

Finally, there are thermal insulation conditions at $r = r_3$ for intumescent fire-retardant coatings wall and edging at $z = L_1 + L_4$

$$\frac{\partial T_3}{\partial r}\Big|_{r=r_3} = 0, \quad S_1 < z \le z_1,$$

$$\frac{\partial T_4}{\partial r}\Big|_{r=r_3} = 0, \quad L_1 < z \le L_4, \qquad (1.4.21)$$

$$\frac{\partial T_4}{\partial z}\Big|_{z=L_1+L_4} = 0, \quad r_3 < r \le r_4,$$

where z_1 can be derived from the formula (1.4.5).

Here and below, α is the heat transfer coefficient; δ is the heat transfer coefficient on the outer surface edging; φ is porous. Subscripts and superscripts: w is external heating surface; (1), (2), (3) subscripts are different rates of the heat flow from the gas phase in Fig. 1.10; *ac* is asbestos cement; *ok* is edging; *no* is no injection.

Design Procedure, Tests, and Initial Data. The experimental temperature dependencies of intumescent fire-retardant coatings mass loss, degree of expansion, and density are taken from [77, 78] and given in Table 1.2. The thermal and physical characteristics of intumescent fire-retardant coatings (SGK-1) found in papers [76, 77] are presented in Table 1.3. Thermophysical coefficients impenetrable steel depending on the temperature are known [81] and are shown in Table 1.4 The formulas for coefficients of thermal conductivity $\lambda_{i,\|}$, $\lambda_{i,\perp}$, $i = 2, 3$ in (1.4.4) are taken from [8].

The boundary problem (1.4.1)–(1.4.3), (1.4.6)–(1.4.8) with the initial and boundary conditions (1.4.9)–(1.4.12), (1.4.14)–(1.4.21) was solved numerically using the locally one decomposition method [82]. We used the implicit, totally stable monotonic difference scheme [45] with total approximation error O $\left(\tau + \sum_{i=1}^{3} H_{L_i}^2 + H_{r_3}^2 + H_{r_4}^2 + H_{ok}^2\right]$, $H_{ok} = (r_4 - r_3)/(N_{ok} - 1)$, where N_{ok} is the number of edging nodes different in r. H_{L_i}, $i = 1, 2, 3$ are the spatial steps along the coordinate z; H_{r_3}, H_{r_4}, H_{ok} are spatial steps along the coordinate r; τ is the time step.

Table 1.2 Structural characteristics of intumescent fire-retardant coatings: relative change in mass, degree of expansion and density of the frame in relation to temperature

T, K	293	350	400	450	500	550	600
$\bar{m}_*$	0.999	0.99	0.98	0.97	0.94	0.88	0.84
θ	0	0.2	1.0	8.5	10	11.3	13
ρ_T, kg/m^3	900	860	525	75	61	60	43
T, K	650	700	750	800	900	1000	1200
$\bar{m}_*$	0.8	0.77	0.72	0.68	0.55	0.54	0.54
θ	16	17	17.4	17.8	18.5	19	19
ρ_T, kg/m^3	44	45	45	45	49	50	50

Table 1.3 Thermal-physical constants of intumescent fire-retardant coatings in depending on temperature

T, K	293	350	400	450	500	550	600
c_{pT}, kJ/(kg K)	0.98	0.99	1.0	1.0	1.15	1.15	1.1
λ_T, W/(m K)	1.0	0.82	0.22	0.23	0.35	0.5	0.7
T, K	650	700	750	800	900	1000	1200
c_{pT}, kJ/(kg K)	1.1	1.05	1.02	1.0	0.99	0.98	0.98
λ_T, W/(m K)	0.83	0.93	0.94	0.95	0.96	0.97	0.97

Table 1.4 Thermal-physical constants of steel in depending on temperature

T, K	293	473	673	873	1073	1273
c_{pm}, kJ/(kg K)	503	510	520	550	600	640
λ_{2m}, W/(m K)	13	14	16	18	21	24

The numerical method was tested for the basic option. The calculation was made under otherwise equal input data for different spatial steps: $H_{L_1} = 2.5 \times 10^{-3}$ m, $H_{L_2} = 0.5 \times 10^{-4}$ m, $H_{L_3,0} = 0.25 \times 10^{-4}$ m, $H_{r_3} = 0.5 \times 10^{-3}$ m, $H_{ok} = 0.5 \times 10^{-3}$ m, $h_{pi} = 2 \cdot H_{L_i}$, $i = 1, 2, 3$, $h_{q_1} = 2 \cdot H_{r_3}$, $h_{q_2} = 2 \cdot H_{ok}$, $h_{xi} = H_{L_i}/2$, $i = 1, 2, 3$, $h_{y_1} = H_{r_3}/2$, $h_{y_2} = H_{ok}/2$, $h_{ui} = H_{L_i}/4$, $i = 1, 2, 3$, $h_{v_1} = H_{r_3}/4$, $h_{v_2} = H_{ok}/4$. The calculation shows that the highest temperature and its gradient are reached in the near-surface layer of the third area (intumescent fire-retardant coatings) due to contact with the heat flow. Consequently, intumescent fire-retardant coatings temperatures were recorded in depth at different times. In all options, the problem was solved at a variable time step chosen on the assumption that prescribed accuracy was equal for all spatial steps. The difference $\delta = \max(\delta_{T_3})$ of relative temperature error dropped by the end of thermal exposure $t = t_k$ and reached $\delta_1 = 9.1\%$, $\delta_2 = 5.2\%$, $\delta_3 = 2.7\%$. The calculation results are presented below for spatial steps $h_{xi} = H_{L_i}/2$, $i = 1, 2, 3$, $h_{y_1} = H_{r_3}/2$, $h_{y_2} = H_{ok}/2$.

The formulas for calculating $(\rho c_p)_{ef}$, f, φ_T and $z_1(t, r)$ given in (1.4.4) and (1.4.5) are taken from [78], while ones given λ_R in (1.4.4) and γ (1.4.13) are taken from [4]. The thermophysical constants for asbestos cement are given in [83], and air is in [84]. Thermokinetic constants for SGK-1 obtained in [76], and presented and approved in the article [77].

The findings presented below were obtained at $T_0 = 293$ K, $(\alpha/c_p)_0 = 0.2$ kg/(m^2 s), $\delta = 100$ W/(m^2 K) $(\alpha/c_p)^{(i)} = 0.01$ kg/(m^2 s), $h^{(i)} = 10^6$ J/kg, $i = 1$, 3, $h^{(2)} = 1.2 \times 10^6$ J/kg, $(\alpha/c_p)^{(2)} = 0.025$ kg/(m^2 s), $T_e = 1200$ K, $L_1 = 5 \times 10^{-2}$ m, $L_2 = 5 \times 10^{-3}$ m, $L_3 = 2 \times 10^{-3}$ m, $L_4 = 2.5 \times 10^{-2}$ m, $\lambda_g = 0.0253$ W/(m K), $d_0 = 4 \times 10^{-4}$ m, $r_1 = 0.25 \cdot r_3$, $r_2 = 0.75 \cdot r_3$, $t_k = 20$ min, $r_3 = 2.5 \times 10^{-2}$ m, $r_4 = 3 \times 10^{-2}$ m, $\sigma = 5.67 \cdot$ W/ (m^2 K^4), $\rho_{ac} = 1800$ kg/m^3, $(c_p)_{ac} = 837$ J/(kg K), $\lambda_{ac} = 0.49$ W/(m K),

$Q = 1.2 \times 10^6$ J/kg, $E = 7.4 \times 10^4$ J/mole, $k = 1.3 \times 10^8$ s^{-1}, $\rho_0 = 900$ kg/m^3, $\rho_{ok} = \rho_{ac}$, $b_1 = 965.5$, $b_2 = 0.0735$, $(c_p)_{ok} = (c_p)_{ac}$, $\lambda_{ok} = \lambda_{ac}$, $N_{ok} = 11$, $\varepsilon_e = 0.9$, $\gamma = 0.19$, $\varepsilon_w = \varepsilon_R = 0.7$.

Results of Numerical Solutions and Analysis. At first, following [78, 80], we considered the quasi-one-dimensional heating process of the three-layer material (without a heat flow along r). Figure 1.11a shows the temperature dependence of intumescent fire-retardant coating surface (T_{3w}) (solid lines).

Figure 1.11b shows the temperature dependence of permeable porous steel (T_{2S_1}) (dashed curves) for $t = t_k$ in the absence of $(Q = 0$—curves under No. $1)$ and the presence of $(Q \neq 0$—curves under No. $2)$ of intumescent fire-retardant coatings decomposition heat.

As can be seen from Fig. 1.11, the difference in temperatures of the protected material (porous steel) may reach 130 K. This result is mainly associated with heat emission from the exothermic reaction when intumescent fire-retardant coating is thermally decomposed.

Now, let us consider the initial (two-dimensional) heating problem. Figure 1.12a presents the temperature dependence of intumescent fire-retardant coating surface (T_{3w}) (solid curves 1, 2).

In Fig. 1.12b (T_{2S_1}) dashed curves 1, 2 stand for $t = t_k$. Curves 1 correspond to $Q = 0$, while curves 2 to $Q \neq 0$. As can be seen from Figs. 1.11 and 1.12, there are qualitative and especially significant quantitative differences at $Q \neq 0$ in the temperature behavior of the composite material (layer 2). This can be explained by heat flowing along the transverse coordinate (r) and uneven heating of intumescent fire-retardant coatings on the part of the external heat flow: $q_w^{(i)} \ll q_w^{(2)} + q_R$, $i = 1, 3$.

Fig. 1.11 Temperature dependence of intumescent fire-retardant coatings surface —solid curves (**a**)—and porous steel surface—dashed curves (**b**)—for $t = t_k$ without regard to heat flow along the transverse coordinate r. Curve under No. 1—$Q = 0$, curve under 2—$Q \neq 0$

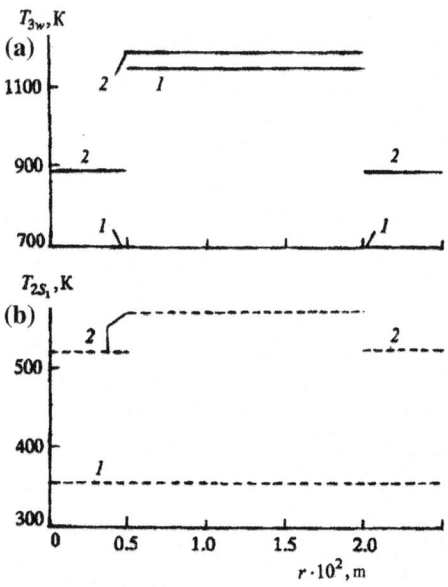

Fig. 1.12 Distribution of temperature at intumescent fire-retardant coatings surface (**a**) and the composite material (**b**) in two-dimensional spatial setting. The symbols are the same as in Fig. 1.10

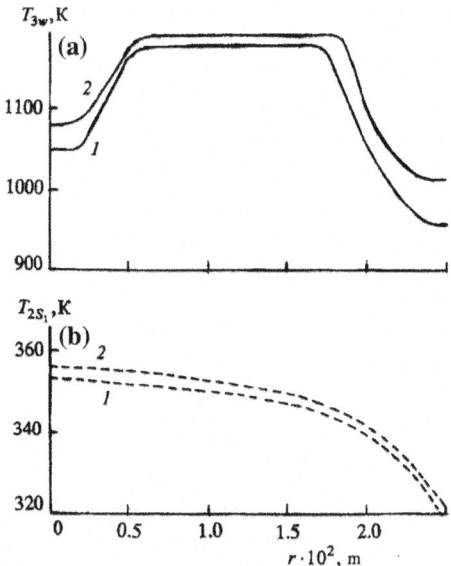

Figure 1.13a shows the time dependence of temperature at junctions of the second and third layers (T_{2S_1}) in the center $(r = r_c)$ of solid steel $(\varphi = 0)$ (dashed curve). The curve was constructed on the basis of quasi-one-dimensional heating conditions (without a heat flow along r) and the initial data of the paper [80]. It follows from Fig. 1.13a that temperature of metal rapidly increases during the first minutes under the effect of the convective heat flow, heating and thermochemical destruction of the intumescent fire-retardant coating layer. This process is accompanied with increasing thickness of the intumescent fire-retardant coating carbonless layer (~ 0.02 m), which has high porosity (~ 0.52) and low thermal conductivity $(\lambda_{\parallel}^{III}/f)$ along the intumescent fire-retardant coating layer with respect to the coordinate s in Eq. (1.4.2), due to 20-fold growth of $f = 1 + \theta$ according to the data given in the third and second lines of Table 1.2. As a result, heating of the protected second layer significantly decreases, while temperature T_{2S_1} of metal becomes stable and does not exceed 550 K during the whole time of heating $t = t_k$. The latter finding is consistent with the finding of [80] achieved by solving the one-dimensional boundary problem of heating of a multilayer body based on a surface fire-protection layer made of SGK-1.

In order to study the effect of the intumescent fire-retardant coatings layer on thermal protection of permeable porous steel, the heating problem was solved in the absence of the intumescent fire-retardant coating layer. The solid curve in Fig. 1.13a was generated by two-dimensional heating of the two-layer body (with asbestos cement used as a permeable material). This illustrates that intumescent fire-retardant coating should be taken into account as lining and heat flowing across the body.

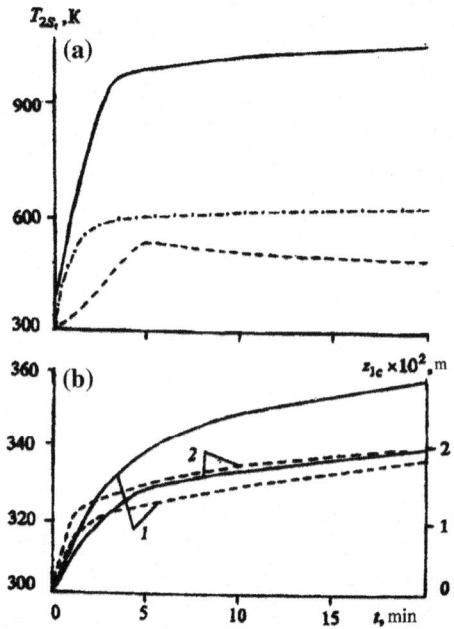

Fig. 1.13 **a** Time dependence of temperature at the junction of the second and third layers in the center $r = r_c$ of solid steel (dashed curve) without regard to heat flow along the coordinate r. Dashed curve is surface temperature of second porous layer for $r = r_c$ at two-dimensional heating. The solid curve stands for the two-dimensional heating of the two-layer model: porous steel—asbestos cement. **b** Time dependence of temperature at the junction of two-dimensional heating of the three-layer tablet (dotted curves) in the center $r = r_c$, dashed curves stand for the coordinates of intumescent fire-retardant coatings external surface at $z_1 = z_{1c}(t, r_c)$. Curve *1* was constructed for $Q = 0$, *2*—$Q \neq 0$

Figure 1.13b presents the solution of the two-dimensional heating problem for the three-layer body in time. The solid curves correspond to T_{2S_1} in the center $r = r_c$, while the dashed curves correspond to the coordinate of the intumescent fire-retardant coating external surface at $z_1 = z_{1c}(t, r_c)$.

Curves *1* were constructed for $Q = 0$, curves *2* for $Q \neq 0$. As can be seen from Fig. 1.13b, temperature of the junction for the option $Q \neq 0$ lies below than one for $Q = 0$. This is associated with an earlier and more rapid increase in thickness of the intumescent fire-retardant coating layer and, therefore, a decrease in heating of the protected second layer.

Chapter 2
Active Thermal Protection

Keywords Active thermal protection · Modeling of heat and mass transfer in a
two-temperature permeable medium · Balance boundary conditions
Porous inert metal materials · Pulsations of the gas stream · Transpiration cooling
Two-phase porous cooling

Hydrodynamic methods based on the injection of a coolant gas into the boundary
layer—the intense heating zone—through porous surfaces are widely used to
protect the head part of flight vehicles and surfaces of power plants against
high-enthalpy and chemically aggressive gas flows [16–24]. The need for active
thermal protection in highly non-isothermal processes arises when a vehicle flies
across dense layers of the atmosphere, as well as in heat and nuclear power plants.
For example, aerodynamic heating significantly increases the temperature of some
structural components of flight vehicles. This may cause destruction of the shell
with a noticeable change in the aerodynamic shapes of vehicles, impair operation of
control devices, and affect the flight range.

The vicinity of the forward critical point is exposed to the heaviest heat loads in a
number of flow conditions. For example, when the Mach number of the incident
airflow $M_\infty = 20$; maximum temperature corresponding to the quasi-equilibrium
state reaches $(6–7) \times 10^3$ K [85]. The known thermal protection systems based on
heat absorption by condensed substances are not effective in coping with heat loads.
In this context, it is reasonable to consider gas injection through the permeable zone.

Due to ever-increasing descent velocities of modern flight vehicles, more
stringent requirements are placed on the thermal protection of structures. Therefore,
the importance of problems addressed in this study is associated with the current
needs of the energy sector and aerospace engineering and confirmed by a great deal
of recent studies devoted to the effects of coolant injection on heat and mass transfer
of protected bodies [16–31].

Studies [20, 57, 86] contain an extensive bibliography of papers studying the
temperature condition of transpiration cooling systems [87–89]. However, as rea-
sonably noted in [86], these findings are not comparable since boundary conditions
are formulated too arbitrarily or inadequately. In Russian and foreign literature [57,
87, 89] devoted to mathematical modeling of heat and mass transfer processes in a

© Springer International Publishing AG, part of Springer Nature 2018   31
A. S. Yakimov, *Thermal Protection Modeling of Hypersonic Flying Apparatus*,
Innovation and Discovery in Russian Science and Engineering,
https://doi.org/10.1007/978-3-319-78217-1_2

two-temperature medium, heat transfer by conduction is usually neglected in the energy conservation equation when a temperature field is calculated. Even when it is taken into account, as in [88], the whole heat flow at the external interface is assumed to be incident on the frame surface (as it would be a single-temperature medium). Moreover, soft boundary condition $\partial^2 T_2/\partial n^2)|_{n=0} = 0$ or thermal insulation condition is established for gas. Research of heat and mass transfer parameters for two types of gas-permeable media [90] shows that balance (i.e., having the meaning of energy conservation laws) boundary laws are invariant to thermophysical properties of a two-temperature porous medium. The extent to which the soft boundary limits can be applied is determined in [90].

The effects of porosity and thermophysical properties of some metals on heat transfer of the models [31] are studied in Sect. 2.1. The heat and mass transfer process in transpiration cooling systems with gas flow pulsations in the one-temperature approximation [30] is modeled in Sect. 2.2. Controllability of heat and mass transfer in thermal protective materials in the two-temperature setting [91] is studied in Sect. 2.3. The heat and mass transfer process in transpiration cooling systems with phase transitions [23] is modeled in Sect. 2.4. The effects of phase transition on heat and mass transfer parameters as well as controllability of heat and mass transfer in thermal protective materials [92] are studied in Sect. 2.5.

2.1   Mathematical Modeling of Heat and Mass Exchange Process in Heat Shielding Material

Porous metal materials are widely used in mechanical engineering, chemical and metallurgical industries, as well as in nuclear and space technologies. Their large thermal stability, high degree of purification (filters) with good permeability, ability to withstand high pressures, corrosion, etc.—all these properties of the porous metal materials ensure their wide applications [2, 17, 29, 93]. The use of the porous metals as heat shielding materials [2, 17, 29, 93] for transpiration cooling makes uniform permeability distribution across the surface one of the most important characteristics [44].

This paper examines the effect of porosity and thermophysical properties of some permeable metals on heat transfer in transpiration cooling systems.

1. **Problem Statement**. To simplify the analysis, we assume that:

   1. The value of mass flow normal to the streamlined plate is much larger than along the plate (Fig. 2.1, $L_1 \ll L_3$).
   2. The body is not damaged when interacting with the gas stream and on there are no homogeneous and heterogeneous chemical reactions and phase transitions on its surface and inside.

Fig. 2.1 Scheme of the body

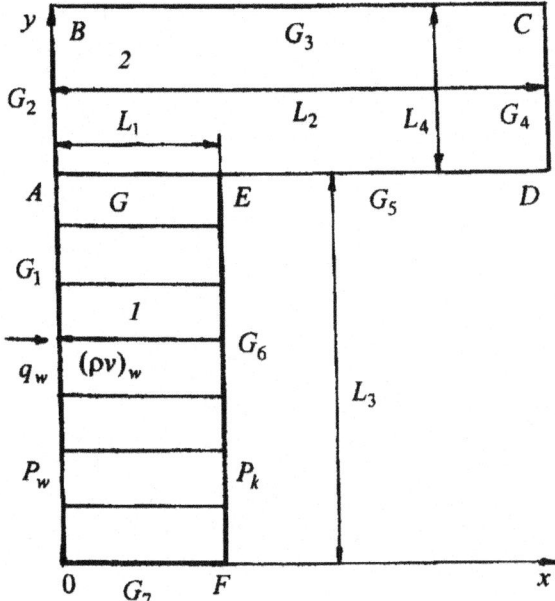

3. The density of the gas phase is determined from the equation of state of ideal gas.
4. The filtrated gas is assumed to be homogeneous with the molecular weight value close to the air mixture.
5. The compositions of the injected gas and the gas in the incoming external flow are the same.
6. The environment is two-temperature, i.e., the gas and condensed phases have different temperatures.
7. The boundary value problem for spatial variables is assumed to be two-dimensional when the heat propagates by thermal conductivity.

In [94], based on the approximate calculated dependences obtained in [95], a formula is given for finding the heat flux on a permeable plate streamlined by a gas stream:

$$\frac{\text{St}}{\text{St}_0} = \frac{\gamma \exp\left(\frac{1-\text{Pr}}{1+\text{Pr}}\gamma\right)}{1 + \frac{2\gamma}{\text{Pr}}\exp\left(\frac{\gamma}{1+\text{Pr}}\right) - \exp\left(\frac{1-\text{Pr}}{1+\text{Pr}}\gamma\right)}, \tag{2.1.1}$$

where $\text{St} = \frac{q_w}{(\rho v)_e(h_e - h_w)}$; $\gamma = \frac{(\rho v)_w}{(\rho v)_e \text{St}_0}$; $\text{St}_0 = \frac{(\alpha/c_p)_0}{(\rho v)_e}$; $\text{Pr} = \left(\frac{c_p \mu}{\lambda}\right)_e$;

St is Stanton number; Pr is Prandtl number; $(\rho v)_w$ is cooler gas flow into the pores of the heat shielding material; c_p is specific heat capacity; μ is dynamic viscosity; h is enthalpy; λ is coefficient of thermal conductivity; α is heat exchange coefficient. Indices: e is the outer limit of the boundary layer; w is heated exterior surface of the body; 0 is no blowing from surface. Here we consider the case when the blown into agent and the main stream are the same.

It is believed that the convective heat flow from the gas phase $q_w(t)$ (Fig. 2.1), acting on heat shielding material for a certain time, is given. Therefore, according to (2.1.1) we have

$$q_w^{(1)} = \frac{(\rho v)_w (h_e - h_w^{(1)}) \exp\left(\frac{1-\text{Pr}}{1+\text{Pr}} b\right)}{1 + \frac{2b}{\text{Pr}} \exp\left(\frac{b}{1+\text{Pr}}\right) - \exp\left(\frac{1-\text{Pr}}{1+\text{Pr}} b\right)}, \qquad (2.1.2)$$

where $b = (\rho v)_w / (\alpha/c_p)_0$; $h_w^{(1)} = c_{p2} T_{1w}$; $c_{p2} = a_1 + 2a_2 T_{2w}$; c_{p2} is specific heat of gas refrigerant at a constant pressure; t is time; T is temperature; a_1, a_2 are constants. Indices: 1 and 2 below are frame and the gas in the 1; (1) and (2) top are area 1 and 2 in Fig. 2.1, 0 is no weakening.

In border zone 2 (to set the boundary conditions), we use the formula of [96], which takes into account the effect of blowing on the heat flux in the form of V. P. Mugalev modifications [4]

$$q_w^{(2)} = (\alpha/c_p)_0 (1 - k_1 b^{k_2})(h_e - h_w^{(2)}), \quad h_w^{(2)} = c_{p2} T_w,$$

where $b = \frac{(\rho v)_w (y=0)}{(\alpha/c_p)_0 (\pi/2 - z)}$, $1 \le z \le 1 + L_4/L_3$, $z = y/L_3$ k_1, k_2 are constants.

The cooling gas consumption will be determined from the simultaneous solution of the stationary equation of continuity and the nonlinear Darcy law [4, 9, 57, 93]. The fact is that at high mass velocities $(\rho v)_w$ (Re > 10) there was observed a turbulent filtering regime [57, 93], which is characterized by a quadratic dependence of the pressure drop rate on the speed.

The task of calculating the heat exchange characteristics is reduced to solving a system of equations of conservation [2, 31]:

in region 1 (see Fig. 2.1)

$$\frac{\partial(\rho_2 v \varphi)}{\partial x} = 0, \qquad (2.1.3)$$

$$\rho_1 c_{p1} \frac{\partial T_1}{\partial t} = \frac{\partial}{\partial x}\left(\lambda_1 \frac{\partial T_1}{\partial x}\right) + \frac{\partial}{\partial y}\left(\lambda_1 \frac{\partial T_1}{\partial y}\right) - A_v(T_1 - T_2), \qquad (2.1.4)$$

$$c_{p2}\left(\rho_2 \varphi \frac{\partial T_2}{\partial t} + \rho_2 v \varphi \frac{\partial T_2}{\partial x}\right) = \frac{\partial}{\partial x}\left(\lambda_2 \varphi \frac{\partial T_2}{\partial x}\right) + \frac{\partial}{\partial y}\left(\lambda_2 \varphi \frac{\partial T_2}{\partial y}\right) + A_v(T_1 - T_2), \qquad (2.1.5)$$

$$A\mu v + B\rho_2 \varphi v|v| = -\frac{\partial P}{\partial x}, \qquad (2.1.6)$$

$$P = \frac{\rho_2 R T_2}{M}, \quad \lambda_2 = \lambda_{2in}\sqrt{\frac{T_2}{T_{in}}}, \quad \mu = \mu_{in}\sqrt{\frac{T_2}{T_{in}}},$$
$$c_{p1} = c_p(T_1)(1-\varphi), \quad \lambda_1 = \lambda(T_1)(1-r\varphi); \tag{2.1.7}$$

in region 2 is the heat equation for a solid steel edging

$$\rho_s c_{ps}\frac{\partial T}{\partial t} = \frac{\partial}{\partial x}\left(\lambda_s\frac{\partial T}{\partial x}\right) + \frac{\partial}{\partial y}\left(\lambda_s\frac{\partial T}{\partial y}\right). \tag{2.1.8}$$

The system of Eqs. (2.1.3)–(2.1.6) and (2.1.8) must be solved with the following initial and boundary conditions:

$$T_i|_{t=0} = T|_{t=0} = T_{in}, \quad i = 1,\ 2; \tag{2.1.9}$$

on the outer heated surface of the porous plate 0A (region 1 in Fig. 2.1, there are some balance boundary conditions [2, 90, 97]

$$(q_w^{(1)} - \varepsilon^{(1)}\sigma, T_{1w}^4)(1-\varphi) = -\lambda_1\left(\frac{\partial T_1}{\partial x}\right)\bigg|_{G_1}, \tag{2.1.10}$$

$$q_w^{(1)}\varphi = -\lambda_2\varphi\left(\frac{\partial T_2}{\partial x}\right)\bigg|_{G_1}; \tag{2.1.11}$$

on line interface AE of regions 1 and 2—the condition of an ideal contact

$$\lambda_1\left(\frac{\partial T_1}{\partial y}\right)\bigg|_{G_-} = \lambda_s\left(\frac{\partial T}{\partial y}\right)\bigg|_{G_+},$$
$$T_1|_{G_-} = T|_{G_+}, \quad T_1|_G = T_2|_G; \tag{2.1.12}$$

on the outer surface of the heated fringing 2

$$q_w^{(2)} - \varepsilon^{(2)}\sigma T_w^4 = -\lambda_s\left(\frac{\partial T}{\partial x}\right)\bigg|_{G_2}; \tag{2.1.13}$$

on the axis of symmetry 0F

$$\left(\frac{\partial T_i}{\partial y}\right)\bigg|_{G_7} = 0, \quad i = 1, 2; \tag{2.1.14}$$

on the surface of a solid steel shell BCD, the conditions of thermal insulation are defined

$$\left(\frac{\partial T}{\partial y}\right)\Big|_{G_3} = 0, \quad \left(\frac{\partial T}{\partial x}\right)\Big|_{G_4} = 0; \tag{2.1.15}$$

on the inner surface *DEF* [97]

$$\left(-\lambda_s \frac{\partial T}{\partial y}\right)\Big|_{G_5} = \delta(T|_{G_5} - T_{in}),$$

$$\left(-\lambda_1 \frac{\partial T_1}{\partial x}\right)\Big|_{G_6} = \delta(T_1|_{G_6} - T_{in}), \tag{2.1.16}$$

$$T_2|_{G_6} = \frac{\delta}{c_{p2}(\rho v)_w}(T_1|_{G_6} - T_{in}) + T_{in}; \tag{2.1.17}$$

on the external and internal surfaces of the region *1*, the equality of pressures in the pores and in the external environment is observed

$$P_w|_{G_1} = P_e, \quad P|_{G_6} = P_{L_1}. \tag{2.1.18}$$

In the given above equations and hereafter the following notation is used: x and y are transverse and longitudinal spatial coordinates; P is pressure; v is gas filtration rate in region *1*; ρ is density; δ is heat transfer coefficient on the inner surface of the plate; A_v is volumetric heat transfer coefficient between the gas and the carcass; R is the universal gas constant; A and B are viscous and inertial coefficients in Darcy law; σ is Stefan–Boltzmann constant; ε is emissivity of the surface heat shielding material; M is molecular mass of air; L_i, $i = \overline{1, 4}$ is thickness and length of the shells *1* and *2* in Fig. 2.1; Pe is Peclet number; Nu is Nusselt number. Indices: *in* is initial value; k is the end of heat exposure; v is bulk quantity; L_1 is the inner side of the body; s is steel; m is molybdenum; t is tungsten.

2. **Calculation Methods and Input Data**. Pressure on the outer surface of the heated permeable body is assumed to be known from the experiment [2, 30] from the longitudinal coordinate y in Table 2.1.

The pressure on the inner "cold" surface of the plate (L_1) is taken as

$$P_{L_1} = kP_{e0}, \tag{2.1.19}$$

Table 2.1 Pressure in the external environment along the streamlined plate

$y \times 10^2$ m	0	0.04375	0.0875	0.175	0.245
$P_e \times 10^{-4}$ Pa	10.0155	9.65	9.3445	9.0139	8.7636
$y \times 10^2$ m	0.35	0.525	0.7	0.875	0.95
$P_e \times 10^{-4}$ Pa	8.1426	7.5116	6.8806	6.2597	6.0093

where k is a constant. It provides the necessary fuel cooler (in particular, the temperature of melting carcass of porous metals is not reached [57, 81]) on a plot of heat exposure from $t = 0$ to $t = t_k$.

Thermophysical factors for continuous steel, molybdenum, and tungsten depending on the temperature are known [81] and are shown in Table 2.2. If the formula of V. I. Odelevsky is used [93], then in the last equation in (2.1.7) for the porous metal: $r = 1.5$ (in the article below, porosity $\varphi \leq 0.44$ was used).

Quasi-stationary equation of continuity $\rho_2 \varphi v = -(\rho v)_w$ (the minus sign is due to the fact that the normal component of the x-coordinate directed into the interior of the body (see Fig. 2.1) and the cooler flows in the opposite direction) in conjunction with the first expression (2.1.7), for the nonlinear Darcy law (2.1.6) and the boundary conditions (2.1.18), can be integrated, and to find the gas flow rate and pressure of 1 [2, 97]:

$$(\rho v)_w(y) = \frac{\left[2B(P_{L_1}^2 - P_w^2)\varphi M D_{L_1}/R + E_{L_1}^2\right]^{0.5} - E_{L_1}}{2BD_{L_1}},$$

$$P(x,y) = \{P_w^2 + 2R(\rho v)_w[B(\rho v)_w D + E]/M\varphi\}^{0.5},$$

where $D(x,y) = \int_0^x T_2(s,y)ds$, $E(x,y) = A \int_0^x \mu T_2(s,y)ds$.

The coefficient of thermal heat transfer A_v between the gas and the carcass is determined by the formula [98]

$$\mathrm{Nu}_v = v_1 \mathrm{Pe}^{v_2}, \quad 0.5 < \mathrm{Pe} < 80, v_2 = 1 - 1.3,$$

where $\mathrm{Nu}_v = A_v l^2/\lambda_2$, $\mathrm{Pe} = (\rho v)_w l c_{p2}/\lambda_2$, $l = B/A$.

The boundary value problems (2.1.4), (2.1.5), (2.1.8)–(2.1.17) are solved numerically using the locally one-dimensional splitting method [82]. An implicit, absolutely stable, monotonous difference scheme with a total approximation error $O(\tau + H_x^2 + H_y^2)$ is used, where H_x is space step along the coordinates x, H_y is a space step along the coordinates y, τ is time step. For the reference variant, the numerical method was tested.

Table 2.2 Thermal characteristics of certain metals on the temperature

металл плотность	T (K)	293	473	673	873	1073
молибден	c_{pm} (J/(kg K))	256	260	267	280	290
12,300 (kg/m³)	λ_m (W/(m K))	140	135	130	125	117
сталь	c_{ps} (J/(kg K))	503	510	520	550	600
7800 (kg/m³)	λ_s (W/(m K))	13	14	16	18	21
вольфрам	c_{pt} (J/(kg K))	132	136	140	144	148
19,350 (kg/m³)	λ_t (W/(m K))	163	156	137	124	116

The calculation was done other things being equal the input data for the different steps in space $H_x = 0.5 \times 10^{-4}$ M, $H_y = 2.4 \times 10^{-4}$ M, $h_{x1} = 2 \cdot H_x$, $h_{x2} = H_x$, $h_{x3} = H_x/2$, $h_{x4} = H_x/4$, $h_{y1} = 2 \cdot H_y$, $h_{y2} = H_y$, $h_{y3} = H_y/2$, $h_{y4} = H_y/4$. Temperature and gas carcass were fixed by the depth of the body at different times. In all cases, the problem was solved with a variable time step, which is chosen from the conditions specified accuracy, the same for all steps in space. The difference $\Delta = \text{MAX}\,[\Delta_{T_1}, \Delta_{T_2}]$ of the relative error in the temperature drops and the time of $t = t_k$ was: $\Delta_1 = 8.3\%$, $\Delta_2 = 4.1\%$, $\Delta_3 = 2.3\%$. The following results were obtained for the calculation steps in space $h_{x3} = H_x/2$, $h_{y3} = H_y/2$.

Thermal characteristics of the air were taken in [60] and the values of A and B in [2, 30]. The following results were obtained when $T_{in} = 293$ K, $(\alpha/c_p)_0 = 0.2$ kg/(s m^2), $\mu_{in} = 1.81 \times 10^{-5}$ kg/(m s), $T_e = 3600$ K, $\lambda_e = 0.782$ W/(m K), $c_{pe} = 4024$ W/(kg K), $h_e = 1.449 \times 10^7$ J/kg, $\mu_e = 1.747 \times 10^{-4}$ kg/(m s), $R = 8.314$ J/(mol K), $M = 29$ kg/кmol, $\sigma = 5.67 \times 10^{-8}$ W/(m^2 K^4), $P_{in} = 10^5$ N/m^2, $L_1 = 2 \times 10^{-3}$ m, $L_2 = 2 \times 10^{-2}$ m, $L_3 = 9.5 \times 10^{-3}$ m, $L_4 = 2 \times 10^{-3}$ m, $\delta = 100$ W/(K m^2), $\lambda_{2in} = 0.0257$ W/(m K), $\lambda_s = 23$ W/(m K), $c_{ps} = 600$ J/(kg K), $\rho_s = 7800$ kg/m^3, $A = 7 \times 10^{10}$ m^{-2}, $B = 10^6$ m^{-1}, $t_k = 10$ s, $a_1 = 965.5$, $a_2 = 0.0735$, $\varepsilon_s^{(1)} = 0.65$, $\varepsilon_m^{(1)} = 0.28$, $\varepsilon_t^{(1)} = 0.39$, $\varepsilon^{(2)} = 0.6$, $v_1 = 0.015$, $v_2 = 1$, $k_1 = 0.285$, $k_2 = 0.165$, $\varphi = 0.36$–0.44.

3. **Rationale for The Approval of Lack of Overflow in Longitudinally Cooler**. In the experiment, [2] used a porous plate is fastened to the holder on both sides (see Fig. 4.3.1 on p. 256 [2]) $L_1 = 2 \times 10^{-3}$ m in thickness and cross section (diameter) $2 \cdot L_3 = 1.9 \times 10^{-2}$ m. A thermocouple was in the middle of the plate soldered on the heated external surface, and the ratio of the plate thickness and length actually vary considerably. Furthermore, the porous plate rests on a solid edging along the length of the two sides, in that the flow of gas to zone 2 of the holder will be absent from $v|_G = 0$ the top and bottom.

Assume that there is gas within the filtration due to changes in temperature porous body along the axis, and using, for example, linear Darcy low

$$v_y = -(\xi/\mu)\partial P/\partial y. \qquad (2.1.20)$$

At $v|_G = 0$ the pressure on the boundary of the time: $P|_G = P_{in} = \text{const}$ and equally, and coefficient of permeability coefficient ξ of the porous medium is given by [93]. Then there will be small gas flow velocity along the plate despite the uneven temperature distribution along the axis y.

To test the assertion, add convective term in the left-hand side of the equation of conservation of energy of the gas phase (2.1.5): $c_{p2}\rho_2\varphi v_y\partial T_2/\partial y$, to evaluate its effect on heat transfer

$$c_{p2}\left[\rho_2\varphi\frac{\partial T_2}{\partial t} + \rho_2\varphi\left(v_x\frac{\partial T_2}{\partial x} + v_y\frac{\partial T_2}{\partial y}\right)\right] = \frac{\partial}{\partial x}\left(\lambda_2\varphi\frac{\partial T_2}{\partial x}\right) + \frac{\partial}{\partial y}\left(\lambda_2\varphi\frac{\partial T_2}{\partial y}\right) + A_v(T_1 - T_2),$$

$$(2.1.21)$$

where v_x, v_y is components of the velocity of the coolant along the axes x, y, respectively, and the density ρ_2 along the axis y is from the first formula of (2.1.7).

Mass conservation Eq. (2.1.3) remains unchanged

$$\partial(\rho_2 v_x\varphi)/\partial x = 0, \qquad\qquad (2.1.22)$$

as in (2.1.22), we neglect the term $\partial\rho_2\varphi v_y/\partial y$, if v_y it is according to the linear Darcy law (2.1.20). It is known [4, 93] that the linear Darcy law (2.1.20) describes the laminar flow of gas, liquid in a porous medium, flow velocity which are orders of magnitude smaller than the velocity of motion in the turbulent flow regime. Then, if you redefine $v_x = v$, then we arrive at the solution of the same problem (2.1.3)–(2.1.6) and (2.1.8)–(2.1.18), where instead of Eq. (2.1.5) it is necessary to use Eq. (2.1.21). To calculate the permeability, (ξ) in Eq. (2.1.20) is taken Karman–Kozeny's formula [93]

$$\xi = \xi_*\varphi^3/(1 - \varphi)^2. \qquad\qquad (2.1.23)$$

Results of Numerical Solutions and Analysis. First, we consider the solution of the support options at the input of the second section and wrap the walls of the tungsten for $\varphi = 0.36$ at $v_y = 0$ ($\xi_* = 0$ in the formula (2.1.23))—the lack of currents cooler along the axis y and at $v_y \neq 0$, $5 \times 10^{-12} \leq \xi_* \leq 5 \times 10^{-10}$ m^{-2}— the presence of this currents. Previously for $\xi_* = 5 \times 10^{-11}$ m^{-2} been obtained for laminar flow regime of gas filtration in modeling heat and mass transfer processes in permeable thermal protection [2, 47].

The numerical calculation was found that a wide range of changes ξ_* in (2.1.23) the temperature of the heated outer surface of the frame does not vary more than 1.2% in relation to the on-surface temperature of the porous wall at $\xi_* = 0$ in the entire length of the axis y. By depth sample x this difference is even smaller.

Table 2.3 shows the dependence of experimental $T_w(0)$ [2] and calculated $T_{1w}(0)$ temperature of the outer surface of the heat shielding material on the gas coolant flow $(\rho v)_w(0)$, the parameter k in Eq. (2.1.19) at the input data in the previous section and porosity $\varphi = 0.36$ for permeable molybdenum. As it can be seen from Table 2.3, the difference of the relative error (Δ) of designed surface temperature from the experimental one is not more than 16.2%.

Table 2.4 shows the dependence of the temperature $T_{1w}(0)$ of the outer surface of the heat shielding material $(\rho v)_w(0)$ for a porous permeable steel and molybdenum at $\varphi = 0.36$. The differences in temperatures of the walls are caused by the thermophysical properties of these metals (apparently, by high thermal conductivity coefficient of molybdenum).

Table 2.3 Relative error of the estimated temperature of the outer molybdenum surface of the carcass in relation to the flow gas cooler

$(\rho v)_w(0)$ (kg/(m^2 s))	0.4	0.6	0.8	1.0	1.2
$T_w(0)$ (K)	720	522	396	354	301
$T_{1w}(0)$, (K)	603	471	417	373	318
Δ (%)	16.2	9.8	5.3	5.4	5.6

Table 2.4 Dependence of the calculated temperature of the external surface heat protective material on the gas flow rate of the cooling

пористый материал	$(\rho v)_w(0)$ (kg/(m^2 s))	0.4	0.6	0.8	1.0	1.2
сталь	$T_{1w}(0)$ (K)	795	566	468	408	365
молибден	$T_{1w}(0)$ (K)	603	471	417	373	318

Furthermore, structural analysis of surfaces made demonstrated [2] that the amount of molybdenum to outputs pores per unit area is greater than that of steel. Porous molybdenum has a fine-grained structure, and it is more developed than that of steel in the specific internal surface at the same porosity. Reducing the particle diameter increases the number of outputs since the entire surface and, thereby, more uniform distribution is given by the coolant.

Consider the effect of porosity on the heat exchange. Figures 2.2, 2.3, and 2.4 show the dependence carcass T_{1w}, temperature, tungsten gas coolant flow rate $(\rho v)_w$ from the heat shielding material surface and q_w convective heat flux from (2.1.2) the gas phase on the longitudinal coordinate y at the end of heat exposure $t = t_k$. Curves *1–3* in Figs. 2.2, 2.3, and 2.4 meet porosity: $\varphi = 0.36, 0.4, 0.44$. From the analysis

Fig. 2.2 Dependence of the outer surface temperature of the frame on the longitudinal coordinate y at time $t = t_k$ porosity for φ: *1*—0.36, *2*—0.4, *3*—0.44

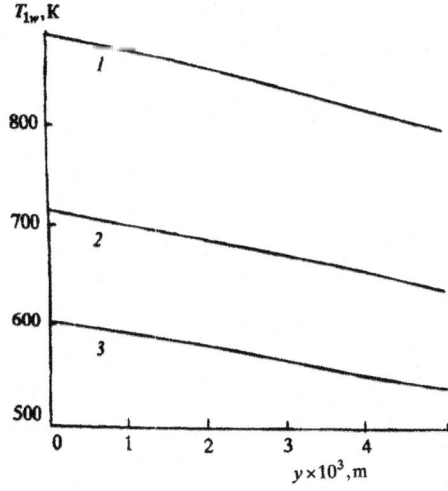

Fig. 2.3 Dependence of the gas flow rate of the cooling of the longitudinal coordinates in a moment of time or $t = t_k$. Symbols are the same as in Fig. 2.2

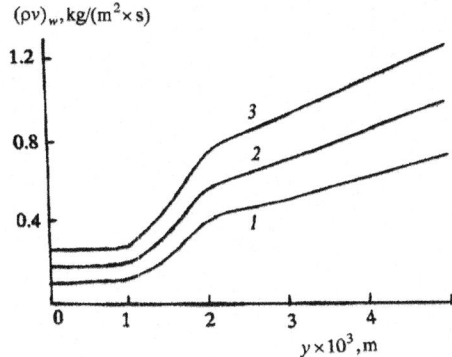

Fig. 2.4 Dependence of convective heat flow from the gas phase of the longitudinal coordinate y at time $t = t_k$. Symbols are the same as in Fig. 2.2

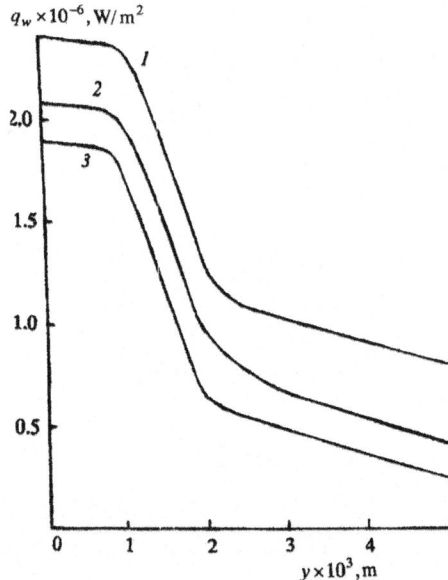

of curves *1–3*, it can be seen that increasing the porosity of the material leads to a decrease in the temperature of perimeter wall.

This result agrees qualitatively with the experimental data [2]. Increasing the porosity results in a more uniform distribution of coolant over the surface and, consequently, in a decrease of heat flow (see Fig. 2.4).

On Fig. 2.5 shows the temperature distribution of the porous steel frame T_1 (solid curves) and gas T_2 (dashed curves) in depth y layer at $x = 0$ and $\varphi = 0.4$ at different moments of time. Figure 2.5 shows that for the balance of the boundary conditions (2.1.10) and (2.1.11) [97] have meaning for conservation laws, $T_{2w} > T_{1w}$. However, as we move deeper into the material quantity T_1 is greater T_2. It related by the sharp increase in the temperature of air at the surface and in a

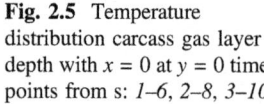

Fig. 2.5 Temperature distribution carcass gas layer depth with $x = 0$ at $y = 0$ time points from s: *1–6, 2–8, 3–10*

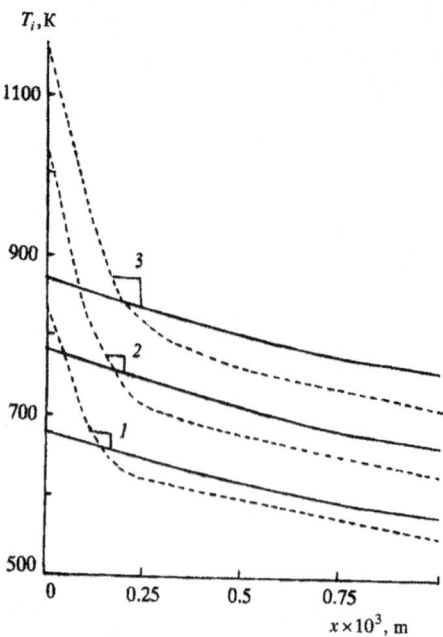

certain neighborhood because of the heat exchange with the external environment and the framework and greater thermal conductivity of the air than the skeleton of the porous steel. Then, the heat transfer process of establishing the condensed phase is warming up faster, since the thermal conductivity of permeable steel two orders of magnitude higher than the thermal conductivity of the gas.

2.2   Modeling of Heat and Mass Exchange Process in Transpiration Cooling System with Gas Flow Pulsation

In real conditions, transpiration cooling systems and thermal protective materials are exposed to low-energy perturbations: wall vibrations, acoustic vibrations, pulsating gas flows [34, 41, 48]. Heat and mass transfer problems in these systems may vary widely [46]. Intensification of heat and mass transfer processes in continuous and permeable media is considered in [30, 31, 34–41]. For example, study [41] revealed that the thermal conductivity coefficient of a porous body increased in the presence of pressure fluctuations on its boundary.

This section is devoted to the effects of gas flow pulsations on the intensity of heat and mass transfer in transpiration cooling systems.

Problem Statement. The comparison of filtration and thermal characteristics of porous materials analyzed in the presence of pulsating and vibrational perturbations [52] demonstrates that dependences of the viscous term in the filtration law and the relative heat transfer function on oscillation intensity and hydrodynamic nature of the heat transfer process are relatively similar to each other. The expression for additional heat transfer q' in a porous body with periodic pulsations of a gas coolant was obtained in [44, 52]:

$$q' = -\frac{c_{p2}\rho_2 W^2 \varphi f}{2\sqrt{2}} \frac{\partial T_1}{\partial x}, \quad f = 2\pi v,$$

where T_1 is the temperature of the one-temperature porous medium (see region 1 in Fig. 2.1); c_{p2}, ρ_2 are respectively the specific heat capacity and the true density of the gas phase of the thermal protective material; W, v are the amplitude of wall pulsations and the frequency of periodic perturbations; x is the transversal spatial coordinate.

The expression the effective viscosity μ_{ef} for the Darcy law can be taken in the following form [61]:

$$\mu_{ef} = \mu[1 + C\cos(tf)], \tag{2.2.1}$$

where C is the dimensionless coefficient $(0 < C \leq 0.2)$.

The varying (pulsating) convective heat flow $q_w(v, t)$ is assumed to act on the thermal protective material for a definite time according to the formula from Sect. 2.1

$$q_w = \frac{(\rho v)_w (h_e - h_w) \exp\left(\frac{1-\mathrm{Pr}}{1+\mathrm{Pr}} b\right)}{1 + \frac{2b}{\mathrm{Pr}} \exp\left(\frac{b}{1+\mathrm{Pr}}\right) - \exp\left(\frac{1-\mathrm{Pr}}{1+\mathrm{Pr}} b\right)}, \tag{2.2.2}$$

$$b = \frac{(\rho v)_w}{(\alpha/c_p)}, \quad \left(\frac{\alpha}{c_p}\right) = \left(\frac{\alpha}{c_p}\right)_0 \left[1 + \frac{U\cos(f\,t)}{(\alpha/c_p)_0}\right],$$

where U is the amplitude of gas flow pulsations.

For the sake of simplicity, we take the assumptions made in Sect. 2.1 and also assume that the medium is one-temperature.

Then, the problem of calculating heat and mass transfer parameters can be mathematically reduced to the system of conservation equations:

in region 1 (see Fig. 2.1)

$$\frac{\partial(\rho_2 v\varphi)}{\partial x} = 0, \tag{2.2.3}$$

$$\rho_1 c_{p1}(1-\varphi)\frac{\partial T_1}{\partial t} + c_{p2}\rho_2 v\varphi\frac{\partial T_1}{\partial x} = \frac{\partial}{\partial x}\left[\lambda_1(1-\varphi)\frac{\partial T_1}{\partial x}\right] + \frac{\partial}{\partial y}\left[\lambda_1(1-\varphi)\frac{\partial T_1}{\partial y}\right],$$

$$(2.2.4)$$

$$A\mu_{ef}v + B\rho_2\varphi v|v| = -\frac{\partial P}{\partial x},\qquad(2.2.5)$$

$$P = \frac{\rho_2 R T_1}{M},\quad \mu = \mu_0\sqrt{\frac{T_1}{T_0}},\quad \lambda_1 = \lambda_1(T_1) + \lambda',$$

$$\lambda' = \frac{\pi\, c_{p2}\rho_2\varphi W^2 v}{\sqrt{2}},\quad c_{p2} = a_1 + 2a_2 T_1;$$

$$(2.2.6)$$

in region 2

$$\rho c_p\frac{\partial T}{\partial t} = \frac{\partial}{\partial x}\left(\lambda\frac{\partial T}{\partial x}\right) + \frac{\partial}{\partial y}\left(\lambda\frac{\partial T}{\partial y}\right).\qquad(2.2.7)$$

The system of Eqs. (2.2.3)–(2.2.5), (2.2.7) must be solved subject to the following initial and boundary conditions:

$$T_1|_{t=0} = T|_{t=0} = T_0;\qquad(2.2.8)$$

on the external heated surface of the porous plate 0A (region 1 in Fig. 2.1)

$$q_w - \varepsilon\sigma T_{1w}^4(1-\varphi) = -\lambda_1(1-\varphi)\left(\frac{\partial T_1}{\partial x}\right)\Big|_{G_1};\qquad(2.2.9)$$

on the symmetry axis 0F

$$\left(\frac{\partial T_1}{\partial y}\right)\Big|_{G_7} = 0.\qquad(2.2.10)$$

The thermal insulation conditions are specified on the surface of the solid steel shell ABCD

$$\left(\frac{\partial T}{\partial x}\right)\Big|_{G_2} = 0,\quad\left(\frac{\partial T}{\partial y}\right)\Big|_{G_3} = 0,\quad\left(\frac{\partial T}{\partial x}\right)\Big|_{G_4} = 0.\qquad(2.2.11)$$

The heat transfer conditions are specified on the internal surface DEF according to Newton's law.

$$\left(-\lambda \frac{\partial T}{\partial y}\right)\bigg|_{G_5} = \delta(T|_{G_5} - T_0),$$

$$\left(-\lambda_1 \frac{\partial T_1}{\partial x}\right)\bigg|_{G_6} = \delta(T_1|_{G_6} - T_0). \tag{2.2.12}$$

The ideal contact condition on conjugation line AE of regions 1 and 2 can be expressed as:

$$\lambda_1(1-\varphi)\left(\frac{\partial T_1}{\partial y}\right)\bigg|_{G_-} = \lambda\left(\frac{\partial T}{\partial y}\right)\bigg|_{G_+}, \quad T_1|_{G_-} = T|_{G_+}. \tag{2.2.13}$$

The pressures in pores and the environment are equal to each other both on external and internal surfaces of region 1.

$$P_w|_{G_1} = P_e, \quad P|_{G_6} = P_k,$$

where φ is the porosity; A, B are respectively the viscosity and inertial coefficients in the Darcy law; I is the intensity of perturbations; a is the speed of sound under normal conditions; index n is nonlinearity. The prime symbol designates fluctuations of heat and transfer parameters.

Numerical Method and Initial Data. The pressure on the external heated surface of the permeable body was assumed to be known from Table 2.1 of Sect. 2.1, being a function of the longitudinal coordinate y. The pressure on the "cold" internal surface of the sphere shell was taken as $P_k = kP_{e0}$ to ensure the necessary coolant flow rate (in particular, to avoid reaching the melting point of the steel shell, i.e., 1600 K) on the thermal exposure area from $t = 0$ to $t = t_k$.

In [2, 30], the expression for the gas coolant flow rate in region 1 was found (see Fig. 2.1):

$$(\rho v)_w(y) = \frac{\left[2(P_k^2 - P_w^2)\varphi M D_{L_1}/R + E_{L_1}^2\right]^{0.5} - E_{L_1}}{2D_{L_1}}, \tag{2.2.14}$$

$$P(x,y) = \{P_w^2 + 2R(\rho v)_w[(\rho v)_w D + E]/M\varphi\}^{0.5},$$

where $D(x,y) = B \int_0^x T_1(s,y)ds, \quad E(x,y) = A \int_0^x \mu_{ef} T_1(s,y)ds.$

The oscillation frequency f can be found by the formula from [2, 46]

$$f = \frac{1}{W}\left(\frac{2I}{\rho_2 a}\right)^{0.5}, \tag{2.2.15}$$

while A and B depending on the intensity of perturbations I are given in [52].

The boundary problem (2.2.4), (2.2.7)–(2.2.13) was solved numerically using the method and the algorithm described in Sect. 2.1.

The thermophysical and structural properties of the porous material were taken from [52, 99] for the sample made of a sintered stainless steel powder and from [84] for air. The results presented below were obtained for $T_0 = 293$ K, $(\alpha/c_p)_0 = 0.2$ kg/(m^2 s), $U = 0.03$ kg/(m^2 s), $\mu_0 = 1.81 \times 10^{-5}$ kg/(m s), $T_e = 3600$ K, $\lambda_e = 0.782$ W/(m K), $c_{pe} = 4024$ J/(kg K), $h_e = 1.449 \times 10^7$ J/kg, $\rho_e = 0.088$ kg/m^3, $\mu_e = 1.747 \times 10^{-4}$ kg/(m s), $W = 10^{-3}$ m, $M = 29$ kg/kmole, $L_1 = 2 \times 10^{-3}$ m, $L_2 = 2 \times 10^{-2}$ m, $\delta = 100$ W/(m^2 K), $a = 340$ m/s, $L_3 = 9.5 \times 10^{-3}$ m, $L_4 = 2 \times 10^{-3}$ m, $\lambda_1 = 2.92 + 4.5 \times 10^{-3} T_1$ W/(m K), $\rho_1 c_{p1} = (1252 + 0.544\,T_1) \times 10^3$ J/(m^3 K), $\varphi = 0.67$, $\lambda = 23$ W/(m K), $c_p = 600$ J/(kg K), $\rho = 7800$ kg/m^3, $I = 0$–0.45 kg/s^3, $t_k = 10$ s, $a_1 = 965.5$, $a_2 = 0.0735$, $\varepsilon = 0.85$, $C = 0.2$, $k = 1.095$.

Discussion of Numerical Solution Results. Figures 2.6, 2.7, and 2.8 present the external surface temperature T_w, the convective heat flow q_w from the gas phase, and the coolant flow rate $(\rho v)_w$ from the surface of the thermal protective material as a function of the longitudinal coordinate y at the end of thermal exposure $t = t_k$.

Lines *1–3* in Figs. 2.6, 2.7, and 2.8 correspond to the intensity of pulsations: $I = 0$, 0.2, 0.4 kg/s^3, i.e., with or without pulsations for heat and mass transfer parameters in the formulas (2.2.1), (2.2.2), (2.2.6).

Figure 2.9 shows how the temperature of the thermal protective material is distributed over the depth of layer x at the forward stagnation point with $y = 0$ and $t = t_k$. The symbols of the lines in Figs. 2.6, 2.7, 2.8, and 2.9 are identical.

The analysis of lines *1–3* (Figs. 2.6, 2.7, 2.8, and 2.9) shows that the periodic perturbations taken into account may both increase or weaken the intensity of heat and mass transfer. This suggests that transpiration cooling systems are sensitive to gas flow pulsations and that the heat and mass transfer process in the thermal protective material can be controlled.

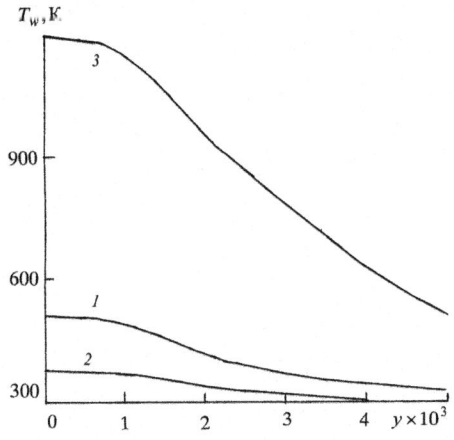

Fig. 2.6 External temperature of the porous plate surface as a function of the longitudinal coordinate y at time $t = t_k$: *1—I* = 0, *2—* 0.2, *3—*0.4

Fig. 2.7 Distribution of the heat flow from the gas phase on the porous surface as a function of the longitudinal coordinate y at time $t = t_k$. The symbols are the same as in Fig. 2.6

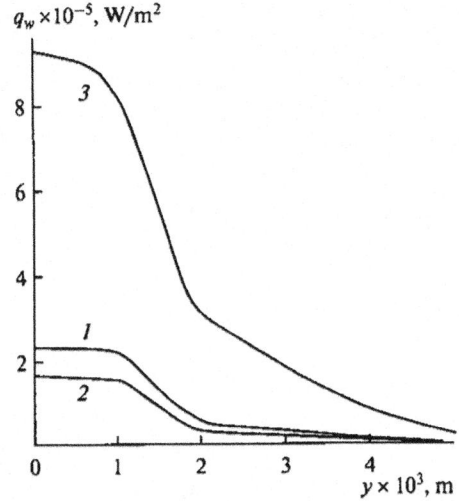

Fig. 2.8 Flow rate of the coolant gas as a function of the longitudinal coordinate y at time $t = t_k$. The symbols are the same as in Fig. 2.6

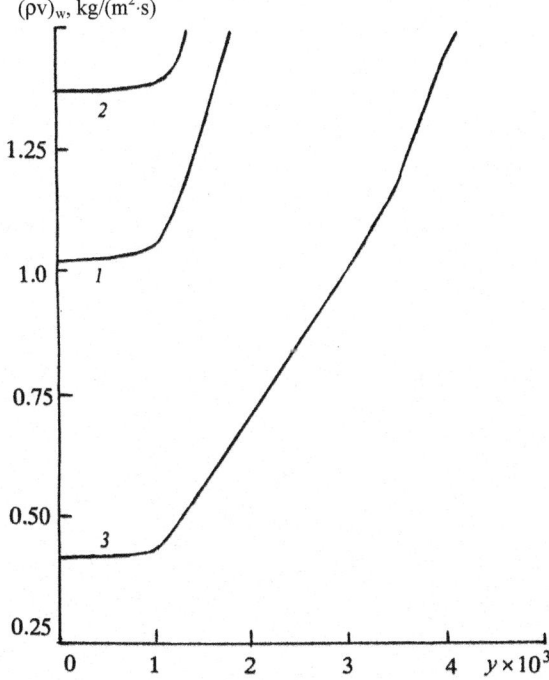

Table 2.5 presents the dependence of the relative heat transfer function $\psi_0 = [q_w^+(0) - q_w^-(0)]/\, q_w^-(0)$ on the intensity of oscillations, where superscripts + and - correspond to the parameters with ($I \neq 0$) and without perturbations

Fig. 2.9 Temperature of
thermal protective material
across the depth of layer x at
$y = 0$ and $t = t_k$. The symbols
are the same as in Fig. 2.6

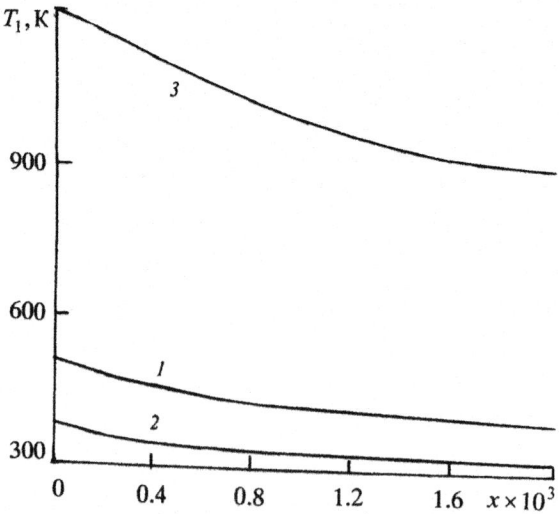

($I = 0$), as well as the surface temperature T_w ($y = 0$) at $t = t_k$ and $b = 5.3$. In
Table 2.5, viscosity α_n and inertial β_n coefficients in the Darcy law (2.2.5) are taken
from Fig. 2 of [52].

As shown in Table 2.5, the behavior of the surface temperature and the relative
heat transfer function is non-monotonic, depending on the intensity of coolant gas
pulsations (2.2.1), (2.2.14).

The latter result is in qualitative agreement with the experimental data [52]. The
non-monotonic behavior of the relative heat transfer function and the viscosity
coefficient α_n in the Darcy law (2.2.5) with growth of the pulsation intensity is
associated with the transition from laminar filtration to transient and turbulent

Table 2.5 Dependence of the surface temperature and the relative heat transfer function on the
intensity of perturbations

No.	I (kg/s³)	$A(\text{m}^{-2})$	B (m^{-1})	ψ_0	$T_w\vert_{y=0}$ (K)
1	0	5.8×10^{10}	1.27×10^6	–	516
2	0.05	4×10^{10}	1.25×10^6	−0.0787	415
3	0.1	3×10^{10}	1.22×10^6	−0.157	363
4	0.15	2.8×10^{10}	1.22×10^6	−0.324	353
5	0.2	3.4×10^{10}	1.24×10^6	−0.391	380
6	0.25	4.6×10^{10}	1.25×10^6	−0.452	419
7	0.3	5.5×10^{10}	1.28×10^6	−0.368	524
8	0.35	6.4×10^{10}	1.33×10^6	−0.341	1028
9	0.4	6.6×10^{10}	1.35×10^6	−0.312	1212
10	0.45	6.8×10^{10}	1.39×10^6	−0.244	1550

filtration. This transition deteriorates permeability of the porous wall [52]. It is reasonable to use a two-temperature permeable medium in the transpiration cooling system in order to reach a quantitative agreement with the experiment [4, 97].

2.3 Modeling of Heat and Mass Transfer Process of Transpiration Cooling Systems Under Exposure to Small Energy Perturbations

Transpiration cooling systems are widely used in engineering applications: evaporation systems, filters, thermal protection elements of flight vehicles [2, 6, 20, 41, 44]. The operation of such systems may be accompanied by some perturbations, such as acoustic oscillations, wall vibrations, pressure fluctuations, and turbulent noises. Depending on the type of perturbations, oscillation amplitude and frequency, thermal and filtration parameters of porous materials may be distorted [44]. For example, in study [41], the thermal conductivity coefficient of a porous body increased in the presence of pressure fluctuations on its boundary.

The purpose of this paper is a theoretical study of transpiration cooling systems in the presence of low-energy perturbations and to compare the obtained findings with the known data.

This section studies the effects of gas flow pulsations on the intensity of heat and mass transfer in transpiration cooling systems with regard to a two-temperature media when gas and condensed phases have different temperatures.

In this case, it is necessary to solve the system of equations from Sect. 2.1: (2.1.3)–(2.1.6), (2.1.8)–(2.1.18), where the expression for $q_w(t, v)$ is known from (2.2.2), μ_{ef} is given in (2.2.1), value A and B are in Eq. (2.1.16) at Table 2.5, and the formula for the thermal conductivity coefficient λ_1 in (2.1.7) is taken as:

$$\lambda_1 = \lambda_1(T_1) + \lambda', \lambda' = \pi c_{p2} \rho_2 \varphi W^2 v / \sqrt{2},$$
$$\lambda_1(T_1) = 2.92 + 4.5 \times 10^{-3} \cdot T_1. \tag{2.3.1}$$

The results presented below are based on the input data of § 2.1, § 2.2, and $U = 0.03$ kg/(s m^2), $W = 10^{-3}$ m, $I = 0$–0.5 kg/ s^3, $t_k = 10$ s, $a = 340$ m/s, $\varepsilon = 0.85$, $C = 0.2$.

Discussion of Numerical Solution Results. Table 2.6 shows the calculated T_{1w} (0) and experimental T_w (0) external surface temperatures of the thermal protective material depending on the coolant gas flow rate, parameter k in the formula (2.1.19), and porosity $\varphi = 0.36$ without gas flow pulsations ($I = 0$). As shown in Table 2.6, the difference in relative errors (Δ) between calculated and experimental temperatures does not exceed 11.9%.

Figures 2.10 and 2.11 present the external temperature of the frame surface T_{1w} and the coolant flow rate $(\rho v)_w$ as a function of the longitudinal coordinate y at $\varphi = 0.67$ at the end time of thermal exposure $t = t_k$. Lines 1–4 in Figs. 2.10 and

Table 2.6 Relative error of the surface temperature a function of the coolant flow rate and parameter k

k	$(\rho v)_w$ (0) (kg/(m² s))	T_w (0) (K)	T_{1w} (0) (K)	Δ (%)
1.072	0.2	1224	1370	11.9
1.092	0.4	1044	968	7.2
1.105	0.6	792	763	3.7
1.122	0.8	648	589	9.1

Fig. 2.10 Dependence of the outer surface temperature of the frame on the longitudinal coordinate y at time $t = t_k$: 1 —$I = 0$, 2—0.1, 3—0.3, 4—0.5

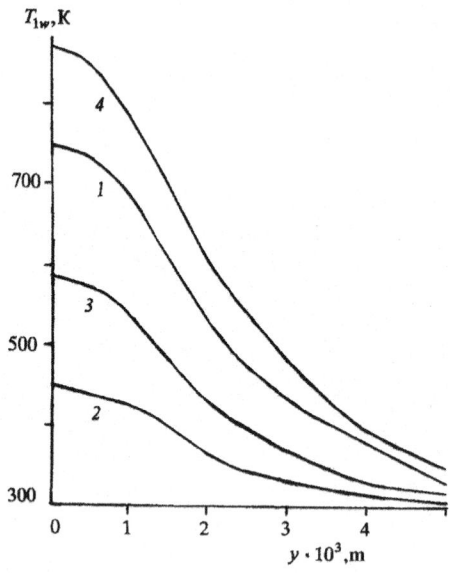

2.11 correspond to the intensity of pulsations: $I = 0$, 0.1, 0.3, 0.5, i.e., for cases without or with pulsations for heat and mass transfer parameters in the formulas (2.2.1), (2.2.2), (2.3.1).

Analysis of lines 1–4 (Figs. 2.10 and 2.11) shows that periodic perturbations taken into account may both increase or weaken the intensity of heat and mass transfer. This suggests that transpiration cooling systems are sensitive to pulsations of the coolant gas flow and that the heat transfer process in the thermal protective material can be controlled.

Table 2.7 presents the temperature of the external surface of the porous steel frame T_{1w} as a function of the coolant flow rate at $\varphi = 0.4$, $v = 0$ and $v = 5.2$ s^{-1}. The values T_{1w} (0) at $x = y = 0$ correspond to the calculation results T_w (0), while the values taken from [100] to the experimental results. The difference in the relative error between the numerical solution and the experiment does not exceed 18.2% at $v = 0$ and 17.1% at $v = 5.2$ s^{-1}.

Fig. 2.11 Dependence of the cooling gas flow from the longitudinal coordinate y at time $t = t_k$. The notation is the same as in Fig. 2.10

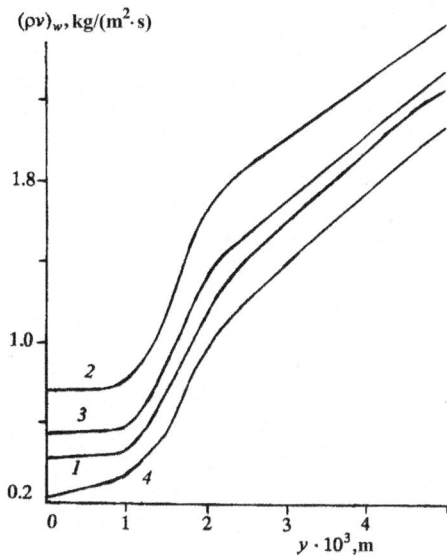

Table 2.7 Relative error of the estimated temperature of the external surface of the frame in relation to the flow of the cooling gas without and with pulsations

v (s^{-1})	0				5.2			
$(\rho v)_w$ (0) (kg/(m^2 s))	0.4	0.8	1.2	1.4	0.4	0.8	1.2	1.4
T_w (0) (K)	1080	666	504	450	900	594	450	360
T_{1w} (0) (K)	933	578	412	373	752	492	375	339
Δ (%)	13.6	13.2	18.2	17.1	16.4	17.1	16.6	5.8

2.4 Modeling of Heat and Mass Transfer of Process of Systems Porous Cooling with Phase Transitions

It is known [4, 20, 100–102] that the most effective method of thermal protection is cooling by water evaporating inside a heated permeable material. Liquid has better thermal characteristics than gas: a high specific heat capacity, large evaporation heat, the possibility to reach low temperatures, and a small specific volume of a liquid coolant.

Heat and mass transfer processes with phase transitions play an important role in the state-of-the-art technology, including thermal protection, drying, metallurgy, etc. Problems of heat transfer with phase transitions for permeable bodies with a substance injected through a porous surface should be regarded as conjugate problems [9, 102]. The most complex issue in the mathematical formulation of these problems is to correctly write the boundary conditions on the phase transition boundary [102].

This section presents a theoretical model of a two-phase transpiration cooling system, compares the calculated data with the experiment, and evaluates the extent to which the mathematical model is applicable.

Problem Statement. A porous plate (Fig. 2.12) with the thickness L_1 is exposed to an external heat flow. Water with the initial temperature T_0 is used as a coolant. Water is forced through permeable wall I–II under the effect of the differential pressure $P_{L_1} - P_0$. As water moves inside the porous plate, it adsorbs heat and its temperature increases. As the temperature increases, the liquid pressure drops. When pressure and temperature of the liquid reach the saturation point, its phase transition occurs.

Evaporation takes place in the plate at the distance x_* from the external heated surface. The differential pressure on the surface of the phase boundary caused by interfacial tension is assumed to be small as compared to the differential pressure across the plate. The resulting steam absorbs heat in the region from the evaporation zone to the external surface and flows out into the environment in the superheated state. As the two-phase mixture moves, the pressure decreases and the mixture temperature has no time to fall after the saturation temperature.

As a result, the mixture remains overheated. An increasing share of vapor in the two-phase flow decreases the intensity of heat transfer. Due to the constant volumetric heat generation, this results in a gradual increase in the temperature of the porous material.

The thermodynamic equilibrium condition of phase transition determines the position of the phase boundary and closes the problem:

Fig. 2.12 Diagram of the flow around a body

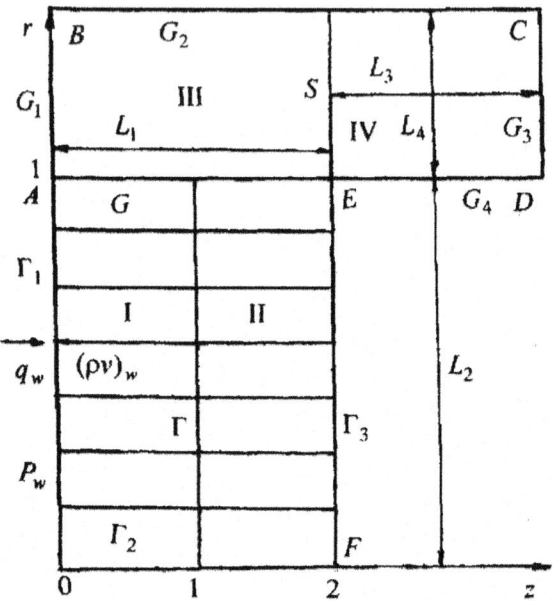

$$x = x_*, \quad P_* = P_*(T_*). \tag{2.4.1}$$

At the moving phase boundary, $x_* = x_*(P_*, T_*)$, the Stephan condition [102] can be written, i.e., the heat balance condition taking into account the absorbed latent heat of the phase transition Q. It is also assumed that temperatures of both phases are equal to each other, and each of them is equal to the phase transition temperature T_* (the condition of phase transition equilibrium):

$$T_2^I|_{x_*} = T_2^{II}|_{x_*} = T_* = \text{const.} \tag{2.4.2}$$

For the sake of simplicity, we assume that:

(1) the Reynolds number in the incident high-entropy airflow is sufficiently high ($\text{Re}_\infty \gg 1$), and a boundary layer was formed in the vicinity of the body surface;
(2) the thermal protective material is a two-temperature medium, i.e., the liquid–vapor and condensed phases have different temperatures;
(3) the mass flow rate at a normal to the plate in flow is significantly greater than that the mass flow rate along this plate (Fig. 2.12, $L_1 \ll L_2$);
(4) the permeable medium I–II (Fig. 2.12) consists of three components: inert frame, water, and superheated vapor;
(5) the density of the liquid–vapor phase is determined from the Clapeyron–Mendeleev equation;
(6) when pressure and temperature of the liquid reach the saturation point, its phase transition occurs;
(7) the two-phase flow in the evaporation zone is homogeneous;
(8) the capillary pressure caused by interfacial tension forces in the evaporation zone is negligible as compared to the total pressure drop across the plate;
(9) the boundary problem in spatial variables is assumed to be two-dimensional.

Since the boundary problem is two-dimensional, and the boundary x_* from (2.4.1) should be preliminarily found, it is reasonable to introduce the transformation of the initial spatial coordinates x and y

along the x-axis
in region I: $z = x/x_*$, $z \in [0, 1]$, $0 \le x \le x_*$,
in region II: $z = x/L_1 + 1, z \in (1, 2], x_* < x \le L_1$,
along the y-axis

$$r = y/L_2, \quad 0 \le y \le L_2 + L_4, \ r \in [0, 1 + L_4/L_2].$$

Then in the new coordinate system r, z:

$$\frac{\partial}{\partial x} = \frac{\partial}{\partial z}\frac{\partial z}{\partial x} + \frac{\partial}{\partial r}\frac{\partial r}{\partial x}, \quad \frac{\partial}{\partial y} = \frac{\partial}{\partial z}\frac{\partial z}{\partial y} + \frac{\partial}{\partial r}\frac{\partial r}{\partial y}, \quad \frac{\partial z}{\partial x} = x_*^{-1}, \quad \frac{\partial r}{\partial y} = L_2^{-1},$$

$$\frac{\partial^2}{\partial x^2} = x_*^{-2}\frac{\partial^2}{\partial z^2}, \quad z \in [0,1], \quad \frac{\partial^2}{\partial x^2} = L_1^{-2}\frac{\partial^2}{\partial z^2}, \quad z \in (1,2], \tag{2.4.3}$$

$$\frac{\partial^2}{\partial y^2} = L_2^{-2}\frac{\partial^2}{\partial r^2}, \quad r \in [0,\ 1+L_4/L_2].$$

The problem for calculating heat and mass transfer parameters is reduced to the system of conservation equations [23, 97, 101]. Then, using (2.4.3), we obtain:
in region I–II (see Fig. 2.12)

$$\frac{\partial(\rho_2 v\varphi)}{\partial z} = 0, \tag{2.4.4}$$

$$A\mu^{\mathrm{I}}v + B\rho_2^{\mathrm{I}}\varphi v|v| = -\frac{\partial P}{x_*\partial z}, \quad z \in [0,1], \tag{2.4.5}$$

$$A\mu^{\mathrm{II}}v + B\rho_2^{\mathrm{II}}\varphi v|v| = -\frac{\partial P}{L_1\partial z}, \quad z \in (1,2], \tag{2.4.6}$$

$$P = \frac{\rho_2 R T_2}{M}, \tag{2.4.7}$$

$$c_{p2}\left(\rho_2\varphi\frac{\partial T_2}{\partial t} + \rho_2 v\varphi\frac{\partial T_2}{x_*\partial z}\right)^{\mathrm{I}} = x_*^{-2}\frac{\partial}{\partial z}\left(\lambda_2\varphi\frac{\partial T_2}{\partial z}\right)^{\mathrm{I}}$$

$$+ L_2^{-2}\frac{\partial}{\partial r}\left(\lambda_2\varphi\frac{\partial T_2}{\partial r}\right)^{\mathrm{I}} + A_{v1}(T_1 - T_2)^{\mathrm{I}}, \quad z \in [0,1], \tag{2.4.8}$$

$$c_{p2}\left(\rho_2\varphi\frac{\partial T_2}{\partial t} + \rho_2 v\psi\frac{\partial T_2}{L_1\partial z}\right)^{\mathrm{II}} = L_1^{-?}\frac{\partial}{\partial z}\left(\lambda_2\psi\frac{\partial T_2}{\partial z}\right)^{\mathrm{II}}$$

$$+ L_2^{-2}\frac{\partial}{\partial r}\left(\lambda_2\varphi\frac{\partial T_2}{\partial r}\right)^{\mathrm{II}} + A_{v1}(T_1 - T_2)^{\mathrm{II}}, \quad z \in (1,2], \tag{2.4.9}$$

$$\rho_1 c_{p1}(1-\varphi)\frac{\partial T_1}{\partial t} = L_1^{-2}\frac{\partial}{\partial z}\left[\lambda_1(1-\varphi)\frac{\partial T_1}{\partial z}\right]$$

$$+ L_2^{-2}\frac{\partial}{\partial r}\left[\lambda_1(1-\varphi)\frac{\partial T_1}{\partial r}\right] - A_{v1}(T_1 - T_2), \quad z \in [0,2]; \tag{2.4.10}$$

in region III

$$\left(\rho_2 c_{p2} \varphi \frac{\partial T_2}{\partial t}\right)^{\text{III}} = L_1^{-2} \frac{\partial}{\partial z}\left(\lambda_2 \varphi \frac{\partial T_2}{\partial z}\right)^{\text{III}}$$

$$+ L_2^{-2} \frac{\partial}{\partial r}\left(\lambda_2 \varphi \frac{\partial T_2}{\partial r}\right) + A_{v2}(T_1 - T_2)^{\text{III}}, \quad z \in [0, 2), \tag{2.4.11}$$

$$\left[\rho_1 c_{p1}(1 - \varphi) \frac{\partial T_1}{\partial t}\right]^{\text{III}} = L_1^{-2} \frac{\partial}{\partial z}\left[\lambda_1(1 - \varphi) \frac{\partial T_1}{\partial z}\right]^{\text{III}}$$

$$+ L_2^{-2} \frac{\partial}{\partial r}\left[\lambda_1(1 - \varphi) \frac{\partial T_1}{\partial r}\right]^{\text{III}} - A_{v2}(T_1 - T_2)^{\text{III}}, \quad z \in [0, 2); \tag{2.4.12}$$

in region IV

$$\rho c_p \frac{\partial T}{\partial t} = L_1^{-2} \frac{\partial}{\partial z}\left(\lambda \frac{\partial T}{\partial z}\right) + L_2^{-2} \frac{\partial}{\partial r}\left(\lambda \frac{\partial T}{\partial r}\right),$$

$$z \in [0, \, 2 + L_3/L_1]. \tag{2.4.13}$$

The convective heat flow from the gas phase is assumed to be known (2.1.2) in region I $0 \le r < 1$ from the expression (2.1.2)

$$q_w = \frac{(\rho v)_w (h_e - h_w^{\text{I}}) \exp\left(\frac{1-\text{Pr}}{1+\text{Pr}} b\right)}{1 + \frac{2b}{\text{Pr}} \exp\left(\frac{b}{1+\text{Pr}}\right) - \exp\left(\frac{1-\text{Pr}}{1+\text{Pr}} b\right)}, b = \frac{(\rho v)_w}{(\alpha/c_p)_0}, \tag{2.4.14}$$

$$h_w^j = (c_{p2} T_{1w})^j, \quad c_{p2}^j = a_1 + 2a_2 T_{2w}^j, \quad j = \text{I, III}.$$

In curtain zone III, we use the formula from [96] in the form of Mugalev's modification taking into account the injection effect on the heat flow [4]

$$q_w^{\text{III}} = \left(\frac{\alpha}{c_p}\right)_0 (1 - k_1 b^{k_2})(h_e - h_w^{\text{III}}),$$

$$b = \frac{(\rho v)_w (r = 0)}{(\alpha/c_p)_0 (\pi/2 - r)}, \quad 1 \le r \le 1.5, \tag{2.4.15}$$

where A is the viscosity coefficient in the Darcy law, A_v is the volumetric heat transfer coefficient between the coolant and the frame, B is the inertia coefficient in the Darcy law, L_i, $i = 1, 2, 3, 4$ are the thicknesses and lengths of shells I–IV in Fig. 2.12, M is the molecular mass of the coolant, Nu_v is the Nusselt number, Pe is the Peclet number, Pr is the Prandtl number, Q is the evaporation heat, r is the dimensionless longitudinal spatial coordinate, v is the filtration rate of the coolant in areas I–II, x and y are the transversal and longitudinal spatial coordinates, respectively, z is the dimensionless transversal spatial coordinate, k_1, k_2 are the constants. Subscripts and superscripts: 1 and 2 correspond to the frame and the coolant in region I–II; L_1 is the internal boundary of the wall by thickness; v is the volumetric quantity.

The system of Eqs. (2.4.4)–(2.4.6), (2.4.8)–(2.4.13) must be solved by taking into account the following initial and boundary conditions:

$$T_2^i\big|_{t=0}= T_0, \quad i = \text{I}, \text{II}; \quad T_1\big|_{t=0}= T\big|_{t=0}= T_0. \tag{2.4.16}$$

The following balance boundary conditions [97] are observed on the external heated surface of the porous plate $0A$ (region I in Fig. 2.12):

$$(q_w - \varepsilon\,\sigma T_{1w}^4)(1 - \varphi) = -(1 - \varphi)\left(\lambda_1 \frac{\partial T_1}{x_*\partial z}\right)\bigg|_{\Gamma_1}, \tag{2.4.17}$$

$$q_w\varphi = -\varphi\left(\lambda_2 \frac{\partial T_2}{x_*\partial z}\right)\bigg|_{\Gamma_1}; \tag{2.4.18}$$

on the conjugation line Γ of regions I and II, according to [102]:

$$\varphi\left(\lambda_2 \frac{\partial T_2}{x_*\partial z}\right)^{\text{I}}\bigg|_{\Gamma_-} = \frac{\varphi}{L_1}\left(\lambda_2 \frac{\partial T_2}{\partial z}\right)^{\text{II}}\bigg|_{\Gamma_+} + Q(\rho v)_w\big|_{\Gamma},$$
$$T_2^{\text{I}}\big|_{\Gamma_-} = T_2^{\text{II}}\big|_{\Gamma_+} = T_*, \tag{2.4.19}$$

on the external heated surface of the porous plate AB (region III in Fig, 2.12) similarly to (2.4.17), (2.4.18):

$$(q_w - \varepsilon\,\sigma T_{1w}^4)^{\text{III}}(1 - \varphi) = -(1 - \varphi)\left(\lambda_1 \frac{\partial T_1}{L_1\partial z}\right)^{\text{III}}\bigg|_{G_1}, \tag{2.4.20}$$

$$q_w^{\text{I}}\psi = -\psi\left(\lambda_2 \frac{\partial T_2}{L_1\partial z}\right)^{\text{I}}\bigg|_{G_1}; \tag{2.4.21}$$

on the internal surface DEF [97]:

$$\left(-\lambda \frac{\partial T}{\partial r}\right)\bigg|_{G_4} = \delta(T\big|_{G_4} - T_0),$$
$$\left(-(1 - \varphi)\lambda_1 \frac{\partial T_1}{L_1\partial z}\right)\bigg|_{\Gamma_3} = \delta(T_1\big|_{\Gamma_3} - T_0), \tag{2.4.22}$$
$$T_2\big|_{\Gamma_3} = \frac{\delta}{c_{p2}(\rho v)_w}(T_1\big|_{\Gamma_3} - T_0) + T_0;$$

on the symmetry axis $0F$:

$$\left(\frac{\partial T_i}{\partial r}\right)\bigg|_{\Gamma_2} = 0, \quad i = 1, 2; \tag{2.4.23}$$

the thermal insulation conditions are specified on the surface of the solid steel shell *ABCD*:

$$\left(\frac{\partial T}{\partial z}\right)\bigg|_{G_i} = 0, \quad i = 1, 3, \left(\frac{\partial T}{\partial r}\right)\bigg|_{G_2} = 0; \tag{2.4.24}$$

the ideal contact condition on the conjugation line *AE* of regions I–II and III:

$$\left(\lambda_1 \frac{\partial T_1}{L_2 \partial r}\right)^{\mathrm{I}}\bigg|_{G_-} = \left(\lambda_1 \frac{\partial T_1}{L_2 \partial r}\right)^{\mathrm{III}}\bigg|_{G_+}, \quad T_1^{\mathrm{I}}\big|_{G_-} = T_1^{\mathrm{III}}\big|_{G_+}, \quad T_1\big|_G = T_2\big|_G, \tag{2.4.25}$$

$$(1 - \varphi)\left(\lambda_1 \frac{\partial T_1}{L_1 \partial z}\right)^{\mathrm{III}}\bigg|_{S_-} = \lambda\left(\frac{\partial T}{L_1 \partial z}\right)\bigg|_{S_+}, \quad T_1^{\mathrm{III}}\big|_{S_-} = T\big|_{S_+}; \tag{2.4.26}$$

pressures in pores and the ambient environment are equal to each other on the external and internal surfaces of regions I–II

$$P_w\big|_{\Gamma_1} = P_0, \quad P\big|_{\Gamma_3} = P_{L_1}. \tag{2.4.27}$$

Numerical Method and Initial Data. The pressure on the inside "cold" surface of the sphere shell is expressed as:

$$P_{L_1} = kP_w, \tag{2.4.28}$$

it ensured a necessary flow rate of a coolant (in particular, melting temperature of the steel shell frame was not reached [4]) in the thermal exposure area from $t = 0$ to $t = t_k$.

The coolant flow rate and the pressure in I–II from [23] have the form:

$$(\rho v)_w(r) = \frac{\left[2(P_{L_1}^2 - P_w^2)\varphi M D_{L_1}/(L_1 R) + E_{L_1}^2\right]^{0,5} - E_{L_1}}{2D_{L_1}},$$

$$P(z, r) = \{P_w^2 + 2Rs(\rho v)_w[(\rho v)_w D + E]/(M\varphi)\}^{0,5},$$

$$s = x_*, 0 \leq z \leq 1, s = L_1, 1 < z \leq 2, \tag{2.4.29}$$

$$D(z, r) = B\int_0^z T_2(r, z)dz; \quad (z, r) = A\int_0^z \mu T_2(r, z)dz$$

The volumetric heat transfer coefficient A_v between gas and the frame was determined from the formulas [98]:

$$\mathrm{Nu}_v = \nu_1 \mathrm{Pe}^{\nu_2}, 0.5 < \mathrm{Pe} < 80, \ \nu_2 = 1-1.3,$$

where $\mathrm{Nu}_v = A_v l^2/\lambda_2$, $\mathrm{Pe} = (\rho v)_w l c_{p2}/\lambda_2$, $l = B/A$.

The boundary problem (2.4.8)–(2.4.13), (2.4.16)–(2.4.26) was numerically solved by means of the locally one-dimensional splitploted method [82]. The implicit, absolutely stable, monotone difference scheme was used with the cumulative approximation error O $(\tau + H_z^2 + H_r^2)$, where H_z is the spatial step along the coordinate z, H_r is the spatial step along the coordinate r, τ is the time step. The numerical method was tested for the basic option. With all initial parameters being equal, the calculation was made for different spatial steps $H_z = 10^{-2}$, $H_r = 0.025$, $H_2 = H_z/2$, $H_3 = H_z/4$, $H_4 = H_z/8$, $h_2 = H_r/2$, $h_3 = H_r/4$, $h_4 = H_r/8$. The temperature of the frame and the coolant was recorded across the body depth at different times. In all cases, the problem was solved with a variable time step chosen on the assumption that the prescribed accuracy was equal for all spatial steps. The difference in $\Delta = \mathrm{MAX}[\Delta_{T_1}, \Delta_{T_2}]$ the relative temperature error decreased and at $t = t_k$ reached: $\Delta_1 = 5.7\%$, $\Delta_2 = 2.4\%$, $\Delta_3 = 1.3\%$.

The thermophysical and structural properties of porous material were taken from [97, 99] for the sample made of a sintered stainless steel powder and from [103] for air and water vapor (Tables II, IV, V). The results presented below were obtained for $A = 6 \times 10^9$ m^{-2}, $\delta = 100$ W/(K m^2), $B = 3.27 \times 10^6$ m^{-1}, $h_e = 1.449 \times 10^7$ J/kg, $L_1 = 10^{-3}$ m, $x_{*0} = 10^{-5}$ m, $L_2 = 7 \times 10^{-3}$ m, $L_3 = 2 \times 10^{-2}$ m, $\lambda_1 = 2.92 + 4.5 \times 10^{-3} T_1$ W/(m K), $\rho_1 c_{p1} = (1252 + 0.544 T_1) \times 10^3$ J/(K m^3), $L_4 = 3.5 \times 10^{-3}$ m, $P_w = P_0 = 10^5$ N/m^2, $Q_0 = 2.2 \times 10^6$ J/kg, $a_1 = 965.5$, $a_2 = 0.0735$, $\varepsilon = 0.9$, $\nu_1 = 0.015$, $\nu_2 = 1$, $\varphi = 0.27$, $t_k = 10$ s.

Quantities x_{*0}, Q_0 were preset at $t = 0$. For $t > 0$, x_* was determined from the conditions (2.4.1), (2.4.2) and the tabular data [103]; $Q(T_*)$ is known from Fig. 1.2 [104].

Method for Determination of Two-Phase Cooling Parameters. The experimental study of the two-phase cooling system was carried out in the air plasma jet generated by EDP-104A/50 plasmatron (designed by the Institute of Thermal Physics, Siberian Branch, Russian Academy of Sciences, Novosibirsk) with parameters $T_\infty = 3600$ K, $V_e = 56$ m/s according to the method described in [105]. The model *1* (Fig. 2.13) was made in the form of a truncated cone with permeable insert *2* embedded in its small base. The cooling liquid (water) *4* was fed through the internal volume of the model and the insert toward plasma jet *3*.

The inserts were made of porous steel by sintering threads of stainless steel X10 H10. The porosity was equal to $\varphi = 0.27$, diameter—to 14×10^{-3} m, thickness— to 10^{-3} m.

The liquid flow rate G was controlled by RS and RMZh rate-of-flow meters and maintained constant during the experiment. The pressure in the chamber P_{L_1} was recorded by MO manometer *5*. The wall temperature T_w in the vicinity of the

Fig. 2.13 Schematic of the tested model

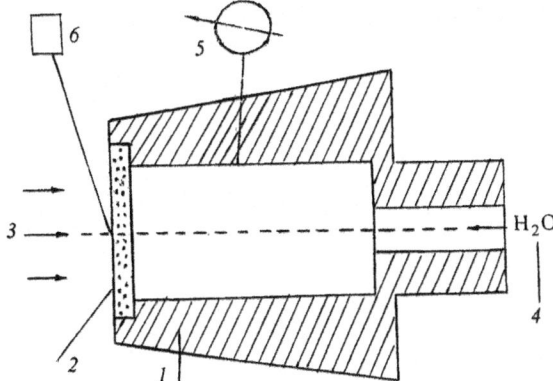

stagnation point was determined by the AKIP-9311 infrared pyrometer *6* as a function of time.

The emissivity $\varepsilon = 0.9$. The signal from the LCD display of the pyrometer was recorded by Canon PowerShot A95 digital camera (frame frequency—15 staff/s). The measured data were processed by a computer. The total measurement errors did not exceed $\delta G \leq 4.5\%$, $\delta P_{L_1} \leq 0.5\%$, $\delta T_w \leq 3\%$.

Experimental Results. Figure 2.14 presents some time dependences of the wall temperature in the vicinity of the stagnation point. Lines *1–4* were obtained for flow rates $(4.12, 2.72, 0.42, 0.39) \times 10^{-3}$ kg/s, respectively. The analysis of the results presented in Fig. 2.14 suggests that the wall temperature changes weakly when liquid is injected through the insert of porous material (lines *1, 2*): the difference in the stationary values of T_w does not exceed 10%.

The vapor cooling mode takes place at low liquid flow rates (lines *3, 4*). Some unstable processes probably associated with the phase transition from areas within the chamber, the porous insert, and the external zone from the direction of the incident plasma, are observed. Unstable processes in the two-phase cooling systems were also observed in studies [101, 105].

Results of the Numerical Solution. Figure 2.15 presents the external temperature of the frame surface T_{1w} (a), the convective heat flow q_w from the gas phase (b), and the coolant flow rate $(\rho v)_w$ (c) from the surface of the thermal protective material as a function of the longitudinal coordinate y (*r*). The solid lines were obtained at $k = 3$ according to the formula (2.4.28), while the dashed lines—at $k = 2.5$. Both of them correspond to *t*: *1*—1 s, *2*—t_k.

The analysis of the lines in Fig. 2.15a–c suggests that the coolant substantially weakens the heat flow according to (2.4.14) at $k = 3$ (see the solid lines in Fig. 2.15b).

Heat absorption takes place both during filtration of a coolant in pores of the thermal protective material and during vapor formation. Moreover, temperature of the porous plate in the vicinity of its joint with the solid steel base $r = 1$ is lower than that at $r = 0$. This result is associated with heat flowing from region I–II to

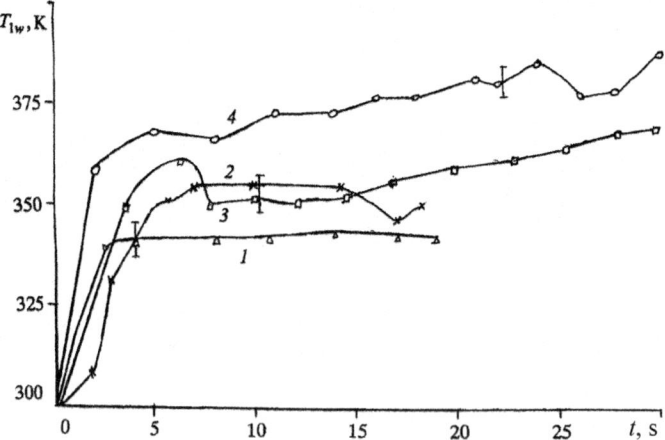

Fig. 2.14 External temperature of the frame external surface as a function of time. The lines were plotted for the following flow rates: 1—4.12×10^{-3}, 2—2.72×10^{-3}, 3—0.42×10^{-3}, 4—0.39×10^{-3} kg/s

region III of the composite body. Figure 2.16 demonstrates the behavior of the frame temperature across the thickness of the thermal protective material x (z) on the symmetry axis at $y = 0$. The symbols in Fig. 2.16 are the same as in Fig. 2.15a.

Figure 2.17 presents the external temperature of the frame surface $T_{1w}(r = 0)$ as a function of time at $k = 2.5$—1 and $k = 3$—2 in the formula (2.4.28). The numerical solution of the problem at $k = 2.5$ is associated with the coolant flow rate $(\rho v)_w = 2$ kg/(s m^2) and the frame surface temperature $T_{1w}(r = 0) = 402$ K. The numerical solution at $k = 3$ is associated with $(\rho v)_w = 2.5$ kg/(s m^2) and $T_{1w}(r = 0) = 390$ K for $t = 7$ s.

The experimental result at $(\rho v)_w = 2.48$ kg/(s m^2) provides $T_w = 369$ K, while at $(\rho v)_w = 2.92$ kg/(s m^2)—$T_w = 361$ K for $t = 7$ s (see Fig. 2.17 and lines 4, 3 in Fig. 2.14). The difference in the relative error between the numerical solution and the experiment is estimated to be respectively 19% and 14% for the coolant flow rate, 9 and 8% for the surface temperature.

2.5 Modeling of Two-Phase Porous Cooling Process at Exposure Low-Energy Perturbations

This section considers the interaction of the two-phase transpiration cooling system with the high-enthalpy pulsating gas flow. It studies the effects of phase transition on heat and mass transfer parameters, as well as controllability of heat and mass transfer in thermal protective materials.

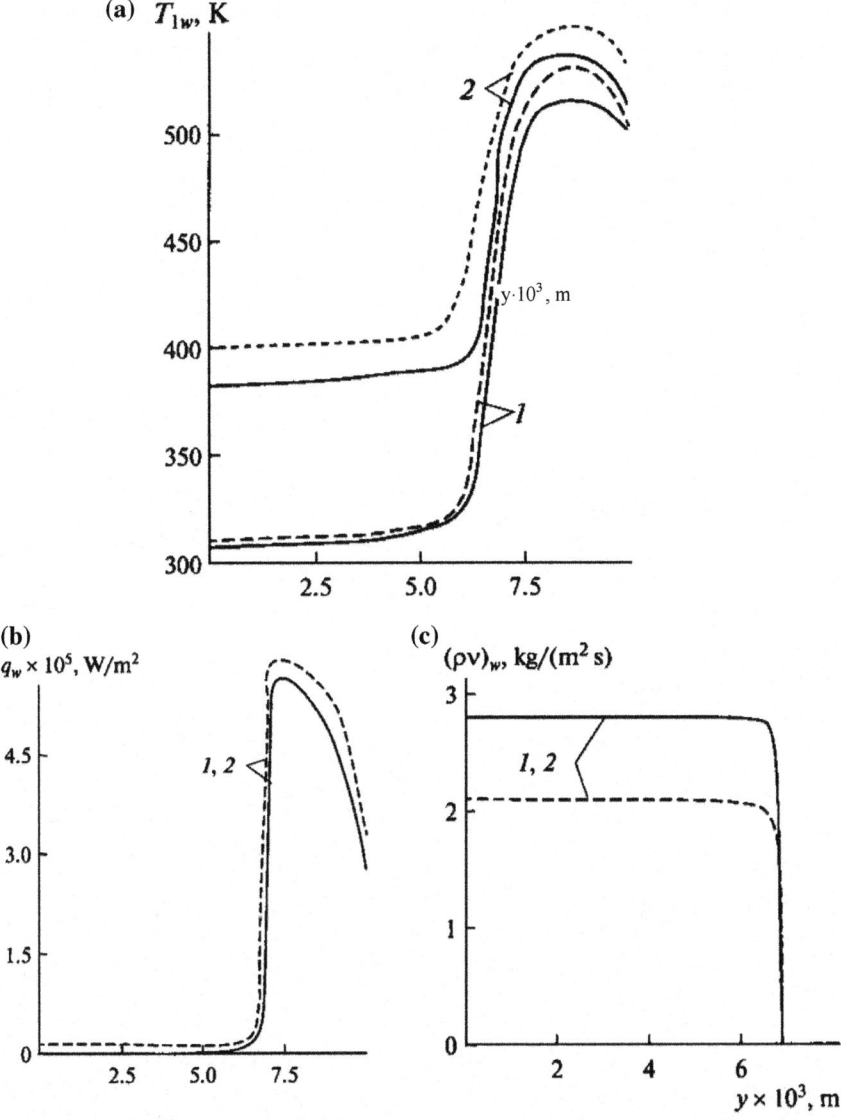

Fig. 2.15 Temperature of the outer surface carcass (**a**) and convective heat flow from the gas phase (**b**), and the coolant flow (**c**) along with a plate, respectively. The solid curves were obtained for $k = 3$, dashed is $k = 2.5$ and t: is 1 s, 2 is t_k

The statement of this problem is similar to the problem in Sect. 2.4, except that the heat flow (2.4.14) is modified by the formula (2.2.2) for the heat transfer coefficient α/c_p:

Fig. 2.16 Temperature of the frame across the thickness of the plate on the symmetry axis *OF*. The symbols are the same as in Fig. 2.9

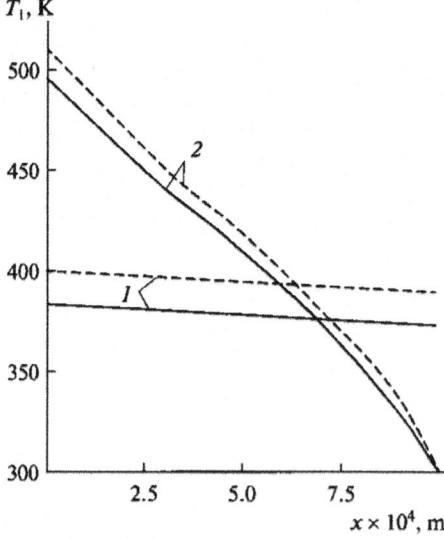

Fig. 2.17 External temperature of the frame surface at $r = 0$ as a function of time for $k = 2.5$—*1* and $k = 3$—*2*

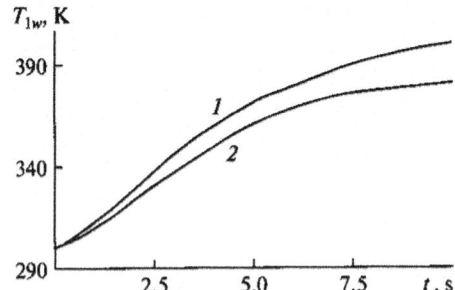

$$\left(\frac{\alpha}{c_p}\right) = \left(\frac{\alpha}{c_p}\right)_0 \left[1 + \frac{W \cos(2\pi v t)}{(\alpha/c_p)_0}\right], \qquad (2.5.1)$$

where W is the amplitude of gas pulsations in the external flow, C is the coefficient of dynamic viscosity in the formula (2.2.1). Eq. (2.4.10) uses the formula (1.2.2) for expressing the thermal conductivity coefficient λ_1.

$$\lambda_1 = \lambda_1(T_1) + \lambda', \quad \lambda' = \pi c_{p5}\rho_5\varphi_5 U^2 v/\sqrt{2}, \qquad (2.5.2)$$

where U is the amplitude of wall pulsations.

The viscosity and inertial coefficient in the nonlinear Darcy law depending on the intensity of pulsations were taken from [52] and given in Table 2.5. The method for determining two-phase cooling parameters is described in [23].

The wall vibrations oriented longitudinally to the incident flow were generated using a vibration exciter based on an electric motor with a gear fixed on an axle. The vibration amplitude $U = 10^{-3}$ m and frequency $v = 0$–20 c^{-1} were prescribed by the geometrical dimensions of the gear and by the motor shaft speed.

Figure 2.18 shows the measured surface temperature of the porous insert in the vicinity of the forward stagnation point of the flow as a function of time. The mass flow rate of the coolant per second is $G = 0.6 \times 10^{-3}$ kg/s. Line *1* was obtained by the experiment without model vibrations, while lines *2*, *3*—with vibrations at the frequency v: 6.8, 12, s^{-1}, respectively.

The analysis of the presented results suggests the attenuation of thermal loads toward the protected wall under the effect of vibrations. Lines *2*, *3* are below line *1* obtained for the wall temperature without perturbations.

Any pulsations and the non-monotonic dependence $T_w(t)$ are not observed.

The flow of the gas–liquid mixture inside the model and through the porous wall is of complex structure and subject to the following conditions: liquid—1, bubble—2, bubble-slug—3, slag—4, slug-annular—5, annular—6, dispersed–annular—7, drop —8, and vapor—9 [24]. Vibrations of the wall (probably due to mixing of the gas–liquid mixture inside the model) eliminate conditions 2–8. This was observed in case of an outflowing gas–liquid mixture when the large opening was used instead of the porous wall [44, 105]. The injection of the liquid and the vapor through the porous wall does not involve unstable processes in the two-phase cooling system.

Fig. 2.18 Experimental temperature of the external heated surface of the body as a function of time. The lines were obtained for the pulsation frequency v: *1*— 0 s^{-1}, *2*—6.8, *3*—12

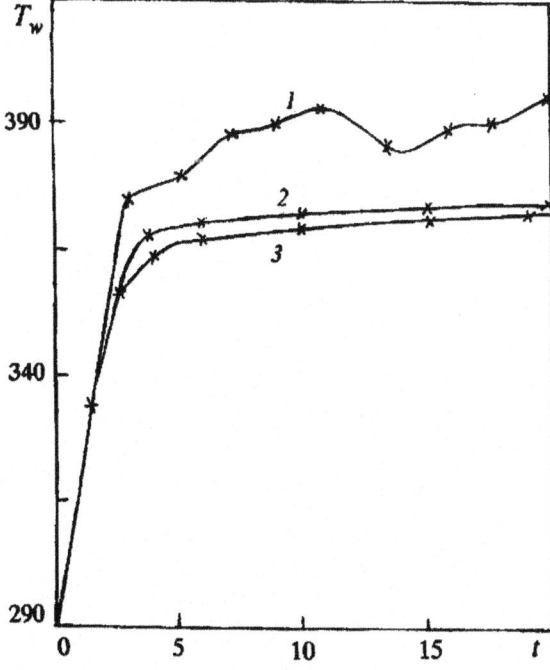

Results of the Numerical Solution. Figure 2.19 presents the external temperature of the frame surface T_{1w} (a), the convective heat flow q_w from the gas phase (b), and the coolant flow rate $(\rho v)_w$ (c) from the surface of the thermal protective material as a function of the longitudinal coordinate y (r). The solid lines were

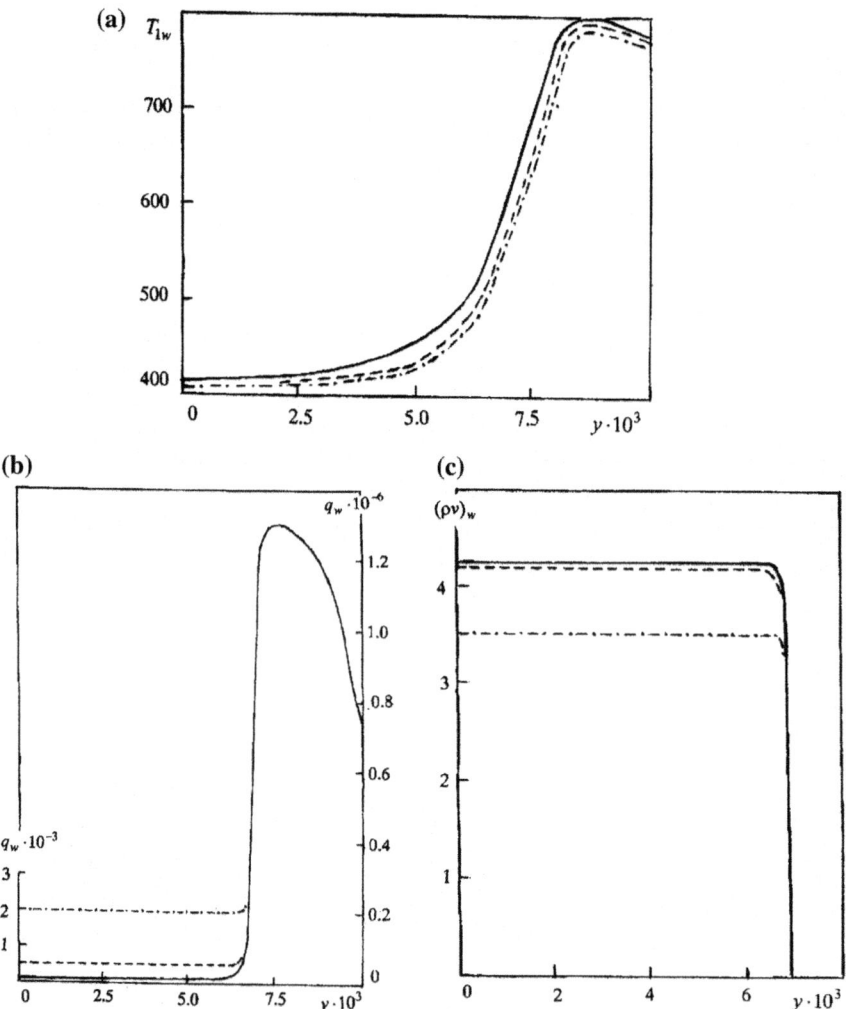

Fig. 2.19 Temperature dependence of the external carcass surface (**a**) along the plate as a function of the coordinate y. The solid lines were obtained at $v = 0$, $W = 0$, $C = 0$, the dashed lines—at $v = 6.8$ s^{-1}, $W \neq 0$, $C \neq 0$, the dot-and-dash lines—at $v = 12$ s^{-1} $W \neq 0$, $C \neq 0$ and $t = t_k$. The convective heat flow from dependence the gas phase (**b**) and the coolant flow rate dependence (**c**) along the plate as a function of the coordinate y. The symbols are the same as in Fig. 2.19a. The coolant flow rate dependence (**c**) along the plate as a function of the coordinate y. The symbols are the same as in Fig. 2.19a

obtained at $v = 0$, $W = 0$, $C = 0$, the dashed lines—at $v = 6.8$ c^{-1}, $W \neq 0$, $C \neq 0$, the dot-and-dash lines—at $v = 12$ s^{-1}, $W \neq 0$, $C \neq 0$, and $t = t_k$.

The analysis of the lines in Fig. 2.19a–c suggests that the coolant substantially weakens the heat flow according to the formula (2.5.1), and a larger flow rate of the coolant $(\rho v)_w$ (see Fig. 2.19c) corresponds to a smaller value of q_w (see Fig. 2.19b). However, as pulsations of the wall of the thermal protective material and the gas flow grow (according to formulas (2.5.1), (2.5.2): $v = 6.8$, 12 s^{-1} $W \neq 0$, $C \neq 0$), the surface temperature decreases (Fig. 2.19a) as a result of the weakening intensity of heat transfer. It should be noted that in curtain zone III (of the uncooled porous edging), the heat flow determined from (2.4.15) and the frame surface temperature are much higher.

The behavior of the frame temperature across the thickness of the thermal protective material x (z) on the symmetry axis at $y = 0$ (lines 1–3) and at the interface with the edging $y = L_2$ (lines $1'$–$3'$) for $t = t_k$ are presented in Fig. 2.20.

The symbols are the same as in Fig. 2.19. As could be expected, the frame temperature T_1 of the cooled plate on the side exposed to the convective flow in region I and partially in region II is lower than that at the interface with region III. This is associated with the heat absorbed during coolant filtration in pores of the thermal protective material and with vapor formation.

On the side of the internal shell surface $y = L_1$ of region II, where water at the initial temperature $T_0 = 293$ K is pumped under pressure, the frame temperature $T_1(y = L_2)$ is lower than T_1 ($y = 0$) as a result of the heat flowing into the "colder" permeable part of plate III and the solid edging IV.

Figure 2.21 presents the external temperature of the frame surface T_{1w} as a function of time at $v = 0$, $W = 0$, $C = 0$–1, $v = 6.8$ s^{-1}, $W \neq 0$, $C \neq 0$–2, $v = 12$ s^{-1}, $W \neq 0$, $C \neq 0$–3. With allowance for pulsations ($v \neq 0$), the surface temperature lines are below the line T_{1w} for $v = 0$, as in the experiment (Fig. 2.18). However, the presence of pulsations can both weaken and stimulate heat transfer (see lines 2, 3, Fig. 2.18) at $t > 6$ s.

The experimental result at $t = t_k$ and $v = 6.8$ s^{-1} is associated with $G = 0.6 \times 10^{-3}$ kg/s, $S = \pi \times L_2^2 = 0.154 \times 10^{-3}$ m^2, $(\rho v)_w = G/S = 3.8$ kg/(m^2 s),

Fig. 2.20 Surface frame temperature across the thickness of plate x on the symmetry axis $0F$ (solid, dashed, and dot-and-dash lines 1–3) and at the interface of regions I–II and III (solid, dashed, and dot-and-dash lines $1'$–$3'$) at $t = t_k$

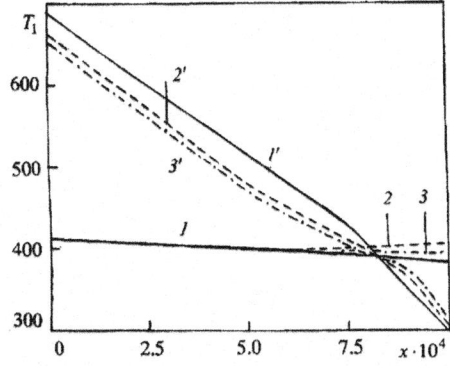

Fig. 2.21 External
temperature of the frame
surface at $y = 0$ as a function
of time. The symbols are the
same as in Fig. 2.18

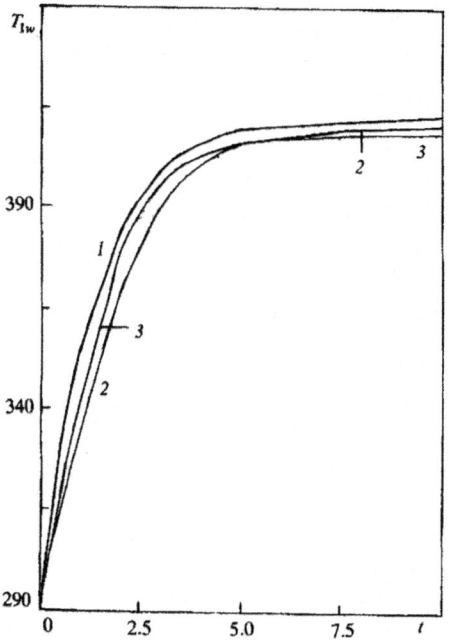

$T_w = 376$ K, while the result of the numerical solution—with $(\rho v)_w = 4.2$ kg/(m^2 s), $T_{1w} = 411$ K. For $t = t_k$ and $v = 12$ s^{-1}, the experiment at the same flow rate $G = 0.6 \times 10^{-3}$ kg/s is associated with $T_w = 370$ K, while the numerical solution—with $(\rho v)_w = 3.5$ kg/(m^2 s), $T_{1w} = 409$ K. The difference in the relative error between the numerical solution and the experiment at $t = t_k$ is estimated to be, respectively, 15 and 8% for the coolant flow rate, 9.3 and 10.5% for the surface temperature.

Chapter 3
Combined Thermal Protection

Keywords Combined thermal protection of hypersonic aircraft
Blunt cone on the sphere · Conjugate heat and mass transfer · Rotation around the
longitudinal axis · Flow around the body at an angle of attack · Heat transfer due to
body rotation · Thermochemical destruction of carbon fiber · Dispersion of carbon
particles · Screening of laser radiation

3.1 Mathematical Modeling of Rotation on Conjugate Heat and Mass Transfer in High-Enthalpy Flow Around a Spherically Blunted Cone at an Angle of Attack

Rotation around the longitudinal axis must be imparted to the body for good
stability in flight. Rotation and injection of a coolant gas change the flow conditions
and thermal effects of incident flow on thermal-protective materials [106, 107]. The
previous studies [42, 106–112] devoted to gas flows around rotating axially symmetric bodies are based on assumptions such as zero angle of attack and an
isothermal shell wall.

In contrast to axially symmetric heating [32], for flow around the body at the
angle of attack [27, 113], the difference in heat flows on leeward and windward
sides may be significant which leads to uneven heating. In order to reduce the
influence of this effect, hypersonic aircraft imparts rotational movement around the
longitudinal axis.

Due to ever-increasing requirements for the stability of aerodynamic parameters,
requires the development of new types of thermal protection [4, 27, 28, 32] based
on the injection of a coolant gas into maximum heat flow zones, rotation, heat flow
control, etc. Research of three-dimensional parabolic and elliptic flows over a wide
range of Mach numbers, Reynolds numbers, and orientation angles for real atmospheric models is required for three-dimensional bodies or bodies that are axially
symmetric at an angle of attack. In order to design and optimize new heat flow
control methods for supersonic flow settings, it is necessary to carry out theoretical

© Springer International Publishing AG, part of Springer Nature 2018   67
A. S. Yakimov, *Thermal Protection Modeling of Hypersonic Flying Apparatus*,
Innovation and Discovery in Russian Science and Engineering,
https://doi.org/10.1007/978-3-319-78217-1_3

and experimental studies in gas and condensed (composite materials) phases taking into account interrelated aerodynamics and heat mass transfer processes. Such a pattern of processes make it necessary to solve aerodynamic problems in a conjugate setting [29]. This allows us to significantly improve the accuracy of determined aerodynamic and heat parameters as compared to studies that separately assess aerodynamics, thermochemical destruction, and motion parameters of the body.

In contrast to [27, 28, 32], this section considers heat and mass transfer and destruction processes associated with the high-enthalpy flow when the body enters the atmosphere at an angle of attack, with regard to rotation and variable injection of a coolant gas along shell lines.

Problem Statement. The studies [9, 114] estimate relaxation times in gas and condensed phases. Based on these estimates, characteristics of conjugate heat and mass transfer are derived from quasi-stationary equations of the spatial boundary layer under different flow conditions. The heat state of the spherically blunted area can be determined by solving the transient energy conservation equation for the porous spherically blunted area and the quasi-stationary equation for filtration rate of a coolant gas in pores within the framework of the one-temperature model.

For a chemical-equilibrium air model, using the assumption of "passivity" and Lewis number equation, the system of equations for the spatial boundary layer in the natural coordinate system associated with external shell surface can be expressed as in [25] (see Fig. 3.1):

$$\frac{\partial}{\partial s}(\rho u r_w) + \frac{\partial}{\partial n}(\rho v r_w) + \frac{\partial}{\partial \eta}(\rho w) = 0, \qquad (3.1.1)$$

Fig. 3.1 Diagram of the flow around a body: 1 is the porous spherically blunted portion, 2 is the conical portion of a body

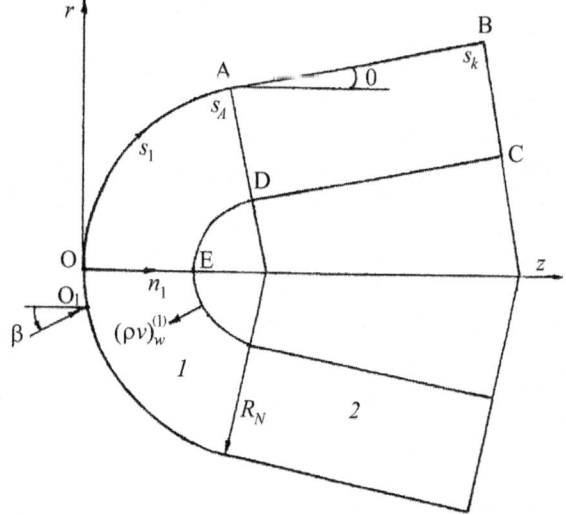

$$\rho\left(u\frac{\partial u}{\partial s}+v\frac{\partial u}{\partial n}+\frac{w}{r_w}\frac{\partial u}{\partial \eta}-\frac{w^2}{r_w}\frac{\partial r_w}{\partial s}\right)=-\frac{\partial P_e}{\partial s}+\frac{\partial}{\partial n}\left(\mu_\Sigma\frac{\partial u}{\partial n}\right), \qquad (3.1.2)$$

$$\rho\left(u\frac{\partial w}{\partial s}+v\frac{\partial w}{\partial n}+\frac{w}{r_w}\frac{\partial w}{\partial \eta}+\frac{uw}{r_w}\frac{\partial r_w}{\partial s}\right)$$
$$=-\frac{1}{r_w}\frac{\partial P_e}{\partial \eta}+\frac{\partial}{\partial n}\left(\mu_\Sigma\frac{\partial w}{\partial n}\right), \qquad (3.1.3)$$

$$\rho\left(u\frac{\partial H}{\partial s}+v\frac{\partial H}{\partial n}+\frac{w}{r_w}\frac{\partial H}{\partial \eta}\right)$$
$$=\frac{\partial}{\partial n}\left\{\frac{\mu_\Sigma}{\mathrm{Pr}_\Sigma}\left[\frac{\partial H}{\partial n}+(\mathrm{Pr}_\Sigma-1)\frac{\partial}{\partial n}\left(\frac{u^2+w^2}{2}\right)\right]\right\}, \qquad (3.1.4)$$

$$P=\rho h(\gamma_{ef}-1)/\gamma_{ef}, \quad P=P_e(s,\eta), \quad H=h+(u^2+w^2)/2 \qquad (3.1.5)$$

$$\mu_\Sigma=\mu+\mu_T, \quad \mathrm{Pr}_\Sigma=\frac{(\mu+\Gamma\mu_T)\,\mathrm{Pr}\,\mathrm{Pr}_T}{\mu\,\mathrm{Pr}_T+\Gamma\mu\mathrm{Pr}}.$$

For the porous spherical shell $(0<s<s_1)$ with one-dimensional filtration of a gas injected toward the normal to the surface in the coordinate system tied to the symmetry axis of the body, we have [28]

$$\frac{\partial[(\rho v)^{(1)}\varphi r_1 H_1]}{\partial n_1}=0, \qquad (3.1.6)$$

$$(\rho c_p)_1(1-\varphi)\frac{\partial T_1}{\partial t}=\frac{1}{r_1 H_1}\left\{\frac{\partial}{\partial n_1}\left[r_1 H_1\lambda_1(1-\varphi)\frac{\partial T_1}{\partial n_1}\right]\right.$$
$$+\frac{\partial}{\partial s_1}\left[\frac{r_1\lambda_1}{H_1}(1-\varphi)\frac{\partial T_1}{\partial s_1}\right]+\frac{\partial}{\partial \eta_1}\left[\frac{H_1\lambda_1}{r_1}(1-\varphi)\frac{\partial T_1}{\partial \eta_1}\right]\right\}$$
$$+c_{pg}^{(1)}(\rho v)_w^{(1)}\frac{r_{1w}}{r_1 H_1}\frac{\partial T_1}{\partial n_1}, \qquad (3.1.7)$$

$$A\mu_1 v^{(1)}+B\rho^{(1)}\varphi v^{(1)}\left|v^{(1)}\right|=-\frac{\partial P}{\partial n_1}, \qquad (3.1.8)$$

$$P=\frac{\rho^{(1)}RT_1}{M}, \quad H_1=\frac{R_N-n_1}{R_N}, \quad \bar{s}=\frac{s_1}{R_N}, \qquad (3.1.9)$$

$$r_1=(R_N-n_1)\sin(\bar{s}), \quad \mu_1\sim\sqrt{T_1}, \quad \lambda_1\sim\sqrt{T_1}, \quad \varphi=\text{const.}$$

For the conical portion, the mass and energy conservation laws in the movable coordinate system $(s_1<s<s_k)$ can be written on the basis of the mathematical models proposed in [4, 14, 28]

$$\rho_c c_{p2} \left(\frac{\partial T_2}{\partial t} - \psi \frac{\partial T_2}{\partial n_1} \right) + c_{pg}^{(2)} G \frac{\partial T_2}{\partial n_1} = \frac{\partial}{\partial n_1} \left(\lambda_2 \frac{\partial T_2}{\partial n_1} \right)$$

$$+ \frac{\partial}{\partial s_1} \left(\lambda_2 \frac{\partial T_2}{\partial s_1} \right) + \frac{1}{r_2^2} \frac{\partial}{\partial \eta_1} \left(\lambda_2 \frac{\partial T_2}{\partial \eta_1} \right) - Q_c \frac{d\rho_c}{dt},$$

$$\tag{3.1.10}$$

$$\frac{d\rho_c}{dt} = \left(\frac{\partial \rho_c}{\partial t} - \psi \frac{\partial \rho_c}{\partial n_1} \right)$$

$$= \begin{cases} -k_c \rho_{c0} \left(\frac{\rho_c - \rho_{c*}}{\rho_{c0}} \right) & \exp \left(-\frac{E_c}{RT_2} \right), & \rho_c > \rho_{c*}, \\ 0, & \rho_c \leq \rho_{c*}, \end{cases}$$

$$\tag{3.1.11}$$

$$G = \int_0^{l_1} \frac{d\rho_c}{\partial t} dn_1, \quad r_2 = (R_N - n_1) \cos \theta + (s_1 - s_A) \sin \theta, \tag{3.1.12}$$

$$l = L - x(t), \quad x(t) = \int_0^t \psi \, d\tau, \quad (\rho v)_{1w} = G_w,$$

$$(\rho v)_w^{(2)} = (\rho v)_{1w} + (\rho v)_{2w} + (\rho v)_{3w}, \quad \psi = \sum_{i=2}^{3} \frac{(\rho v)_{iw}}{\rho_{cw}},$$

In addition, further taken the following assumption: The characteristic linear speed of rotation of the body is much less than the oncoming flow rate

$$\Omega = \frac{\omega R_N}{V_*} \ll 1. \tag{3.1.13}$$

The initial conditions:

$$T_1|_{t=0} = T_2|_{t=0} = T_0, \quad \rho_c|_{t=0} = \rho_{c0}. \tag{3.1.14}$$

The boundary conditions in a gas phase can be written as follows:
at the external interface of the boundary layer at $n \to \infty$

$$u \to u_e(s, \eta), \quad w \to w_e(s, \eta), \quad h \to h_e(s, \eta), \tag{3.1.15}$$

where P_e, u_e, w_e, h_e can be found from the system of Euler equations [115];
at the body surface at $n = 0$

$$u(s, \eta) = 0, \quad w = w_w, \quad v = v_w, \quad (0 < s < s_1). \tag{3.1.16}$$

For body rotation, $w_w = \omega r_w$, where ω is the angular velocity of rotation. For instances without rotation, $w_w = 0$.

The balance conditions $(n = n_1 = 0)$ are identified for the external shell surface [28] at $0 \leq \eta < 2\pi$

$$\frac{\mu}{Pr} \left(\frac{\partial h}{\partial n} \right) \Big|_w - \varepsilon_1 \sigma T_{1w}^4 (1 - \varphi) = -\lambda_1 (1 - \varphi) \left(\frac{\partial T_1}{\partial n_1} \right) \Big|_w , \quad 0 < s < s_A; \quad (3.1.17)$$

$$-\lambda \frac{\partial T_2}{\partial n_1} \Big|_{n_1 = 0 - x(t)} = \frac{\mu}{Pr} \left(\frac{\partial h}{\partial n} \right) \Big|_w - (h_w - h_c) \sum_{i=2}^{3} (\rho v)_{iw} \quad (3.1.18)$$
$$- (\rho v)_{1w} (h_w - h_g) - \varepsilon_2 \sigma T_{2w}^4, \quad s_A \leq s \leq s_k.$$

The following relations can be written for the internal surface of the hemisphere and the conical Newton's condition

$$\lambda_1 (1 - \varphi) \frac{\partial T_1}{\partial n_1} \Big|_{n_1 = L} = \delta (T_{1,L} - T_0), \quad 0 < s_1 < s_A, \quad (3.1.19)$$

$$P_c |_{n_1 = l} = P_{c0}, \quad \lambda \frac{\partial T_2}{\partial n_1} \Big|_{n_1 = \ell} = 0, \quad s_A \leq s_1 \leq s_k, \quad (3.1.20)$$

Perfect contact conditions are used at the sphere-cone interface ring $s = s_1$. The adiabatic condition at $s_1 = s_k$ is

$$\frac{\lambda_1 (1 - \varphi)}{H_1} \frac{\partial T_1}{\partial s_1} \Big|_{s = s_1 - 0} = \lambda_2 \frac{\partial T_2}{\partial s_1} \Big|_{s = s_1 + 0}, \quad (3.1.21)$$

$$T_1 |_{s = s_1 - 0} = T_2 |_{s = s_1 + 0}, \quad \frac{\partial T_2}{\partial s_1} \Big|_{s_1 = s_k} = 0.$$

Pressures in pores and the ambient environment are equal to each other on external and internal surfaces of the spherically blunted area:

$$P_w |_{n_1 = 0} = P_e(s, \eta), \quad P |_{n_1 = L} = P_L. \quad (3.1.22)$$

For the flow with a plane of symmetry:
In the absence of flow symmetry planes:

$$T_1(t, n_1, s_1, \eta_1) = T_1(t, n_1, s_1, \eta_1 + 2\pi), \quad (3.1.23)$$
$$T_2(t, n_1, s_1, \eta_1) = T_2(t, n_1, s_1, \eta_1 + 2\pi).$$

At $s \geq s_1$, the following kinetic pattern of non-equilibrium chemical reactions was considered at the interface of media $(T_{2w} \approx 4000 \, \text{K})$ [9, 54]:

$$C + O_2 \rightarrow CO_2,\ 2C + O_2 \rightarrow 2CO,\ C + O \rightarrow CO,\ C + CO_2 \rightarrow 2CO,$$
$$2O + C \rightarrow O_2 + C,\ 2N + C \rightarrow N_2 + C,\ C \leftrightarrow C_1,\ C \leftrightarrow C_3 \tag{3.1.24}$$

The molar and mass flow rates of chemical reactions (3.1.24) are described in detail in [9, 13]. The mass rate of ablation can be expressed as [13]:

$$(\rho v)_{2w} = \rho_w \left[\left(\frac{m_6}{m_2} - 1 \right) c_{2w} B_1 + \left(2\frac{m_5}{m_2} - 1 \right) c_{2w} B_2 + \right.$$
$$\left. + \left(\frac{m_5}{m_1} - 1 \right) c_{1w} B_3 + \left(2\frac{m_5}{m_6} - 1 \right) c_{6w} B_4 \right],$$

$$(\rho v)_{3w} = \sum_{i=7}^{8} \frac{m_i A_{ci} (P_{ci}^* - P_{ci})}{(2\pi R T_{2w} m_i)^{0,5}},\ i = 7, 8,$$

$$P_{ci}^* = 10^5 \cdot \exp(D_i - E_i/T_{2w}), \tag{3.1.25}$$

$$B_i = k_{iw} \exp(-E_{iw}/RT_{2w}),\ i = \overline{1,4},\ P_{ci} = P_e c_{iw} m_w/m_i,\ i = 7, 8,$$

$$\rho_w = P_e m_w/(RT_{2w}),\ h_w = \sum_{i=1}^{8} h_i c_{iw},\ m_w = \sum_{i=1}^{8} \frac{c_{iw}}{m_i},$$

$$c_g = b_1 + b_2 T,\ h_g = \int_0^T c_g dT.$$

In (3.1.25) and (3.1.26), serial numbers of components correspond to the following order: O, O_2, N, N_2, CO, CO_2, C_1, C_3.

We can write the balance relationships for mass concentrations of components (c_{iw}) using the Fick's law for diffusion flows, as well the analogy of heat and mass transfer processes [4, 9]:

$$J_{iw} + (\rho v)_w^{(2)} c_{iw} = R_{iw},\ i = \overline{1,8},$$
$$J_{iw} = \beta_i (c_{iw} - c_{ie}),\ \beta_i = \alpha/c_p,$$

where α/c_p and β_i are, respectively, the heat and mass transfer coefficients. Disruption products are considered to weakly dilute the air mixture in the boundary layer. This assumption makes it possible to use the setting accepted above for the equations in the boundary layer.

Hereinafter, u, v, w are the components of mass average velocity in the natural coordinate system (s, n, η); Γ is the intermittency coefficient; H, m are, respectively, the total enthalpy and the molecular mass; R_N is the radius of the spherically blunted area; r_w, r_i, $i = 1, 2$, H_1 are the Lame coefficients; h and $(\rho v)_w^{(1)}$ are, respectively, the enthalpy and the flow rate of a coolant gas from the spherically blunted surface; $(\rho v)_w^{(2)}$ is the total ablation from the carbon surface of the conical portion; φ the porosity of the spherically blunted area, L is the shell thickness, θ is the taper angle; β is the angle of attack; n_1 is the normal to surface toward the inside of the shell; ω is the linear velocity of the destructed surface; $x(t)$ is the interface between the gas and condensed phases (burn-up depth); c_{iw} is the mass concentration of the component; E_{iw}, k_{iw}, $i = 1, \ldots, 4$ are, respectively, the activation energy and the pre-exponential factor of the heterogeneous reaction on the shell of the conical portion.

Subscripts e, e0, and w stand for values at the external interface of the boundary layer at the stagnation point and on the body surface; 1, 2 "down"—for of frame and gas on the sphere; g—for the gas phase on the conical portion of the surface; ∞—for values of the incident gas flow at infinity; τ, 0—for characteristics of turbulent transfer and initial conditions; k—for the peripheral region of the shell. Superscripts (1) and (2) correspond to characteristics associated with the flow rate of a coolant on the porous hemisphere and surface chemical reactions on the conical portion of the body, the overline—to non-dimensional parameters. z—to the end time of thermal exposure, ef—to the effective value, c—carbon fiber Reinforced Plastics.

Calculation Method and Initial Data. The system of Eqs. (3.1.1)–(3.1.4), (3.1.6)–(3.1.8), (3.1.10) and (3.1.11), with the initial and boundary conditions (3.1.15)–(3.1.23) was numerically solved. The system of equations for the spatial boundary layer was solved in terms of Dorodnicyn variables, taking into account laminar, transitional, and turbulent regions of the flow. The two-layer model of the turbulent boundary layer was used to describe the turbulent flow [116, 117]. The three-layer algebraic turbulence model accounts for the laminar viscous sublayer, the internal area of the turbulent core described by the Van Driest–Cebeci formula [117] and the external area in which the Spalding formula is used [116]. The transition point is assigned on the assumption that the heat flow reaches its peak on the sonic line of the spherically blunted area under given pressure and stagnation enthalpy values. The intermittency coefficient and transition from the laminar to the turbulent flow were described by the Dhvani-Narasimha formula [118]. For numerical integration, Pr = 0.72, $\text{Pr}_T = 1$. As related to boundary layer conditions, the iterated interpolation method [45] was used to develop combined difference schemes ensuring combination of target values at the boundary of the laminar, transition, and the turbulent core, and accounting for a pattern of change in μ_T across the boundary layer. The described boundary layer model was tested by comparing it with the experimental findings of [119, 120]. The test demonstrated the good performance of this model.

The three-dimensional Eqs. (3.1.7) and (3.1.10) were solved by splitting method [82]. The implicit, absolutely stable, monotone difference schemes were used, with cumulative approximation error $O(\tau + H_{n_1}^2 + + H_s^2 + H_\eta^2)$, where H_{n_1} is the spatial step along the coordinate n_1, H_s is the spatial step along the coordinate s, H_η is the spatial step along the coordinate, τ is the time step.

The results of theoretical [121] and generalized experimental studies [122] were used for testing the processes associated with interaction of high-enthalpy air flows with graphite surfaces.

The quasi-stationary equation of continuity (3.1.6) $\rho_2 \varphi v = -(\rho v)_w \times r_{1w}/(H_1 r_1)$ (the negative sign is explained by the fact that the normal coordinate n_1 is directed into the depth of the body (see Fig. 3.1), and the coolant flows in the opposite direction) can be integrated along with the first equation (3.1.9), the nonlinear Darcy law (3.1.8) and the boundary conditions (3.1.22) to find gas flow rate and pressure in the region 1 [4, 97]:

$$(\rho v)_w^{(1)}(s_1, \eta_1) = \frac{\left[2B\left(P_L^2 - P_w^2\right)\varphi \, MD_L/R + E_L^2\right]^{0.5} - E_L}{2BD_L}, \qquad (3.1.26)$$

$$P(n_1, s_1, \eta_1) = \left\{ P_w^2 + 2R(\rho v)_w^{(1)}\left[B(\rho v)_w^{(1)}D + E\right]/M \, \varphi \right\}^{0.5},$$

where $D(n_1, s_1, \eta_1) = \int_0^{n_1} T_2 \left(\frac{r_{1w}}{r_1 H_1}\right)^2 dy$, $E(n_1, s_1, \eta_1) = A \int_0^{n_1} \mu_1 T_2 \frac{r_{1w}}{r_1 H_1} dy$.

The pressure on the inside "cold" surface of the plate (L) is expressed as:

$$P_L = k P_{e0}, \qquad (3.1.27)$$

where k is a certain constant. It ensured the necessary flow rate of a coolant (in particular, melting temperature of the frame made of porous metals was not reached [4, 57]) in the thermal exposure area from $t = 0$ to $t = t_z$.

Calculations cone flow around a spherically blunted, with the angle of the semi-solution $\theta = 10°$ the flow of chemical equilibrium of air under attack angle $\beta = 5°$ were performed for the following conditions [115], which are coresponsible options $V_\infty = 7000\,\text{m/s}$, $H_\infty = 2.2 \times 10^4\,\text{m}$ $R_N = 0.2\,\text{m}$, $L_0 = 0.02\,\text{m}$. The kinetic constants (3.1.25) of heterogeneous reactions (3.1.24) were taken from [9], the enthalpy of the graphite was calculated by the formula [58]. Effectiveness adiabatic index in the first formula (3.1.5) was determined according to [115]. For carbon material coefficients of thermal conical shell are known from [14], for a porous steel–of [99].

The following results were obtained for: $h_{e0} = 2.07 \cdot 10^7\,\text{J/kg} = 0.34$, $T_0 = 300\,\text{K}$, $M = 29\,\text{kg/kmole}$, $\sigma = 5.67 \times 10^{-8}$ $\text{W/(m}^2 \cdot \text{K}^4)$, $\varepsilon_2 = 0.9$, $P_0 = 10^5\,\text{N/m}^2$, $\rho_{c0} = 1400\,\text{kg/m}^3$, $\rho_{c*} = 1300\,\text{kg/m}^3$, $k_c = 3.15\,10^6\,c^{-1}$, $E_c = 8.38 \times 10^4\,\text{J/mole}$, $Q_c = 1.26\,10^6\,\text{J/kg}$, $t_z = 40\,\text{s}$. Thermal characteristic bluntness porous corresponded porous steel: $\lambda_1 = 2.92 + 4.5 \times 10^{-3} \cdot T_1\,\text{W/(m K)}$, $\rho_1 c_{p1} = (1252 + 0.544 \cdot T_1) \times 10^3\,\text{J/(K kg/m}^3)$, $A = 2.3 \times 10^{11}\,\text{1/kg/m}^3$, $B = 5.7 \times 10^5\,\text{1/m}$ [99, 123], $\varepsilon_1 = 0.8$. The thermal characteristics of the conical body responsible carbon fiber [14] or the impenetrable graphite V-1 [124].

Assessing the Impact of the Rotational Speed on the Boundary Conditions on the Surface of the Body. In order to justify the methodology used solutions of the dual problem, given that the temperature of the body surface changes due to aerodynamic heating and rotation, give estimates of the characteristic times in the boundary layer on the surface of the body and the effect of rotational speed on the boundary conditions Eqs. (3.1.17) and (3.1.18).

According to [13] define the relaxation time of the gas phase

$$t_a = R_N/V_* \approx 0.2\,\mathrm{m}/7000\,\mathrm{m/s} = 2.86 \times 10^{-5}\,\mathrm{s}.$$

The characteristic thermal relaxation time of the solid body on a surface [13] is

$$t_w = \left(\rho\, c_p \lambda\right)_i [(T_{e0} - T_0)/q_*]^2, \quad i = 1, 2.$$

For porous steel [99] from the last indent have at $q_* = 10^7$ W/m^2, $T_{e0} - T_0 = 7905$ K: $(\rho\, c_p \lambda)_1 = 9 \times 10^6$ J^2/(m^4 K^2 s), $t_{w1} = 5.58$ s. For solid graphite V-1 on a conical surface at $(\rho\, c_p \lambda)_2 = 9 \times 10^6$ J^2/(m^4 K^2 s) [124] is received $t_{w2} = 15.9$ s, and for carbon fiber [14] at $\rho c_{p2} \lambda_2 = 3.57 \times 10^6$ J^2/(m^4 K^2 s) have $t_{w3} = 2.21$ s.

We estimate the change in the surface temperature caused by the rotational movement of the body. Suppose that at time t at the point (s, η) in the coordinate system associated with the gas phase, have a surface temperature $T_w(s_1, \eta_1)$. After a period of time Δt due to the rotation of the body to the point (s, η) « will » temperature $T_w(s_1, \eta_1, +up\Delta \eta_1)$. Then

$$|T_w(s_1, \eta_1, \Delta \eta_1) - T_w(s_1, \eta_1)| = |(\partial T_w/\partial \eta_1) \times \Delta \eta_1|.$$

Given that $(\Delta \eta_1) = |\omega \times \Delta t|$ we have

$$|T_w(s_1, \eta_1, +\Delta \eta_1) - T_w(s_1, \eta_1)| = |(\partial T_w/\partial \eta_1 \times \omega \Delta t)|.$$

To assess the value $|\partial T_w/\partial \eta_1|$ of the top take the distribution $T_w(\eta)$ of solutions body flow problem at $\omega = 0$ rad/s. According to the results of calculation of temperature difference between two points on the body surface $\eta = 0$ (leeward side) and $\eta = \pi$ (upwind) at time $t = 40$ s reaches values of 1000 K. We estimate the value $|\partial T_w/\partial \eta_1|$ of the magnitude of $1000/\pi$ K/rad ≈ 318 K/rad. Given that in this study used the angular rotation rate of 0.436–8.73 rad/s and substituting Δt characteristic time t_a to the gas phase, we obtain

$$|(\partial T_w/\partial \eta_1) \times \omega\, t_a| \approx 0.004\,\mathrm{K} \text{ at } \omega = 0.436\,\mathrm{rad/s}$$

$$|(\partial T_w/\partial \eta_1) \times \omega\, t_a| \approx 0.08\,\mathrm{K} \text{ at } \omega = 8.73\,\mathrm{rad/s}.$$

It is clear from these estimates that the characteristic relaxation time of the gas phase much shorter than the thermal relaxation the characteristic time of the solid body on the surface $t_a \ll t_w$; during a change t_a in temperature of the surface (the boundary condition for the gas phase) by the rotational movement is very small compared with the surface temperature $(T_w \approx 2400 - 3500\,\text{K})$. Consequently, the flow in the boundary layer in the framework of this can be regarded as quasi-stationary mathematical formulation of the problem.

It is worth noting that less than ω, the weaker pace of change-ture at the surface due to rotational motion (the characteristic time of change $t_\omega \sim \pi/\omega$).

Discussion of the numerical solution results. Figures 3.2 and 3.3 present the surface temperature of a body along the contour as a function of the longitudinal coordinate $\bar{s}$ in sections of the circumferential coordinate $\eta = 0-\pi$ for composite materials: porous steel and carbon-fiber reinforced polymer (Fig. 3.2), permeable steel and V-1 graphite (Fig. 3.3). Curves 1–4 in Figs. 3.2 and 3.3 correspond to

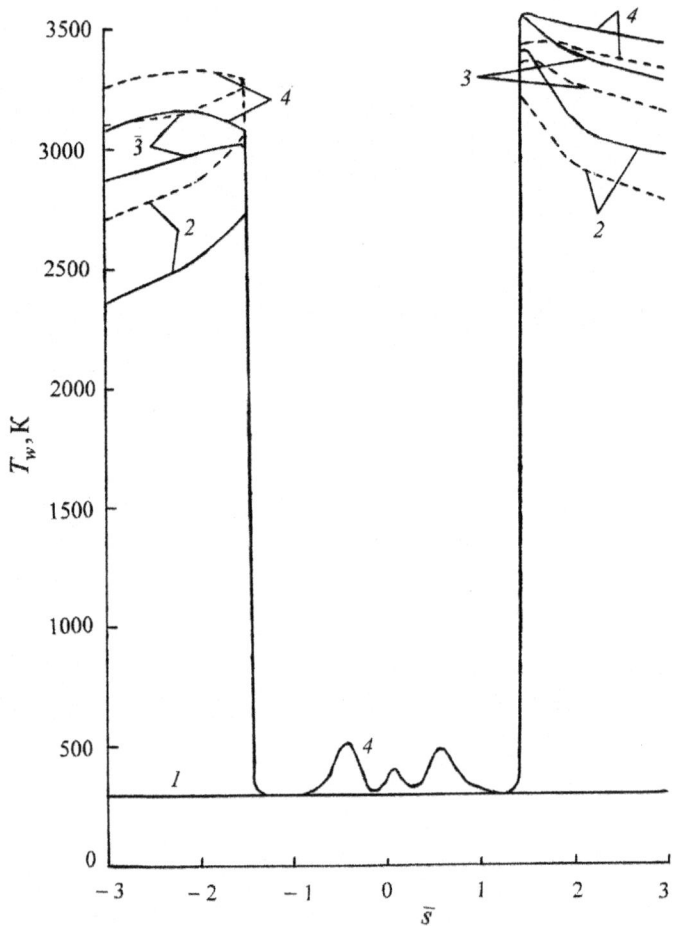

Fig. 3.2 Surface temperature along the contour as a function of the coordinate in the section $\eta = 0-\pi$ for the porous steel—carbon-fiber-reinforced polymer (CFRP) composite body at time points t: 1–0, 2–1, 3–5, 4–t_z

time points t: 1–0, 2–1, 3–5, 4–t_z ($t_z = 40$ s corresponds to the stationary heating process of a body).

The solid curves in Figs. 3.2 and 3.3 represent the case without rotation, while dashed lines are plotted at $\Omega = 0.00027$ ($\omega = 8.727$ rad/s). Figures 3.4, 3.5, 3.6, 3.7 and 3.8 calculations are for $t = t_z$. The curves in Fig. 3.2 correspond to the coolant gas flow rate $(\rho v)_w^{(1)}$ from (3.1.26) under number 1 in Fig. 3.4 determined at $k = 1.06$ from the formula (3.1.27). The surface temperature curves in Fig. 3.3 correspond to $(\rho v)_w^{(1)}$ under number 2 in Fig. 3.4 determined at $k = 1.1$.

In these cases, we can select the distribution of required pressure in the chamber (P_L) for porous spherically blunted metal portions so that the critical temperature

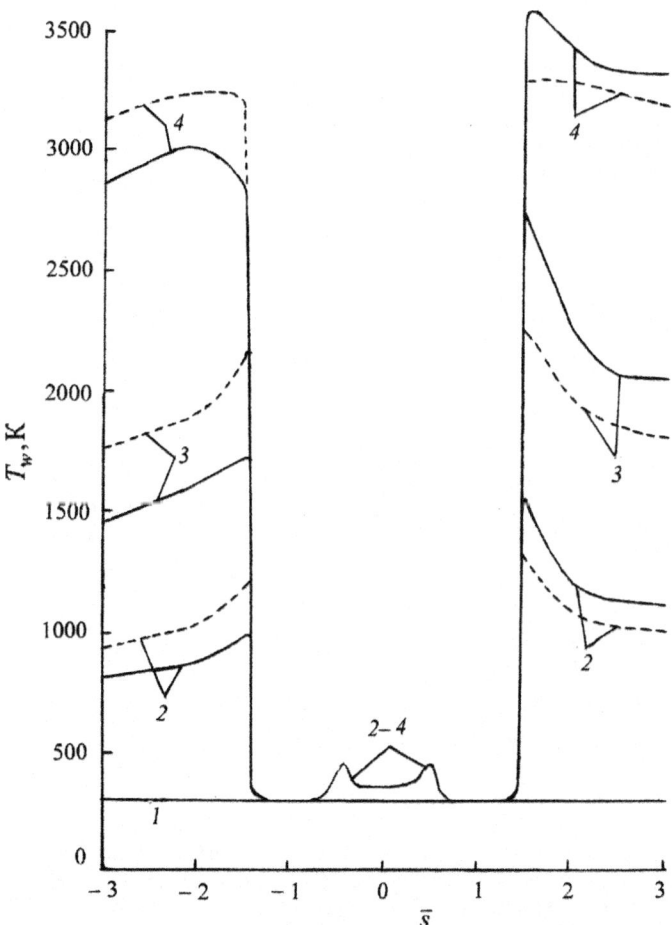

Fig. 3.3 Surface temperature distribution along the contour at $\eta = 0$—π for the permeable steel—V-1 graphite composite body. The symbols are the same as in Fig. 3.2

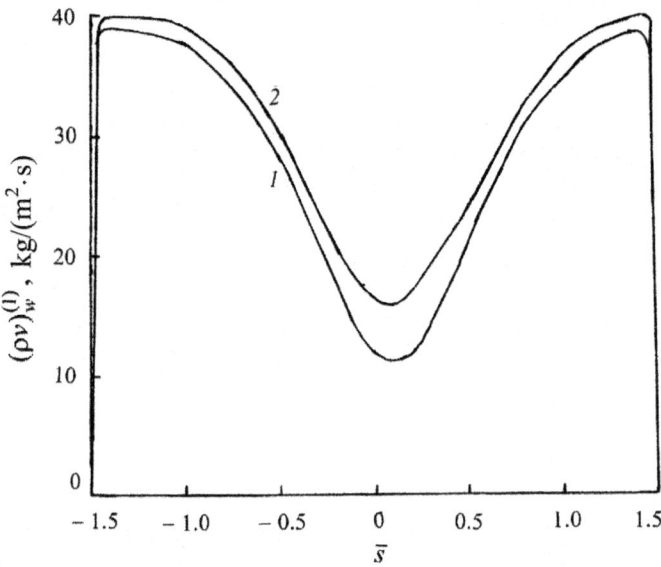

Fig. 3.4 Distribution of the mass flow rate of the coolant gas along the contour on the spherically blunted portion at $t = t_z$. Curve *1* corresponds to the porous steel—CFRP composite body, *2*—to the permeable steel—V-1 graphite composite body

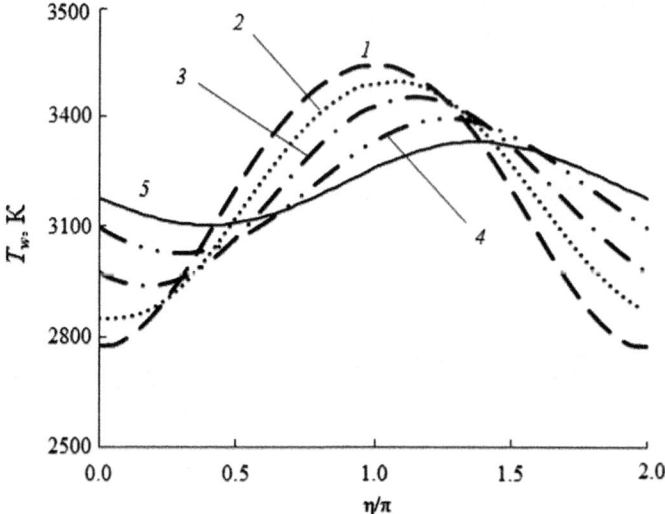

Fig. 3.5 Surface temperature on the conical portion of the body (CFRP) as a function of the circumferential coordinate in the section of the longitudinal coordinate $\bar{s} = 10$. Curves *1, 2,..., 5* correspond to rotational speeds of 0, 25, 100, 250, 500°

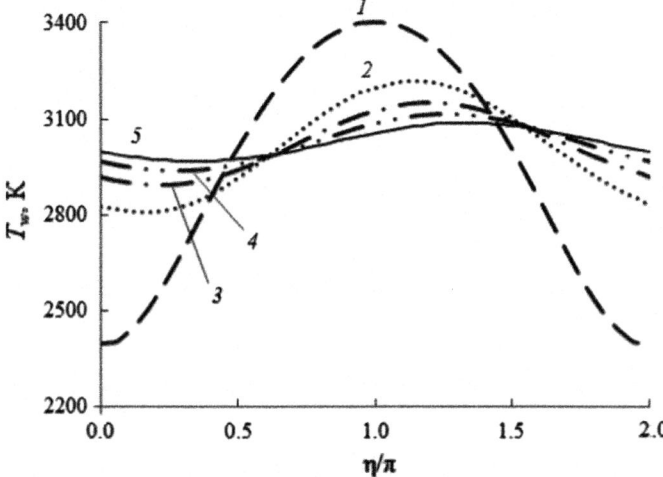

Fig. 3.6 Distribution of surface temperature of the conical body (graphite V-1) in the circumferential coordinate in the cross section $\bar{s} = 10$. Symbols are as in Fig. 3.5

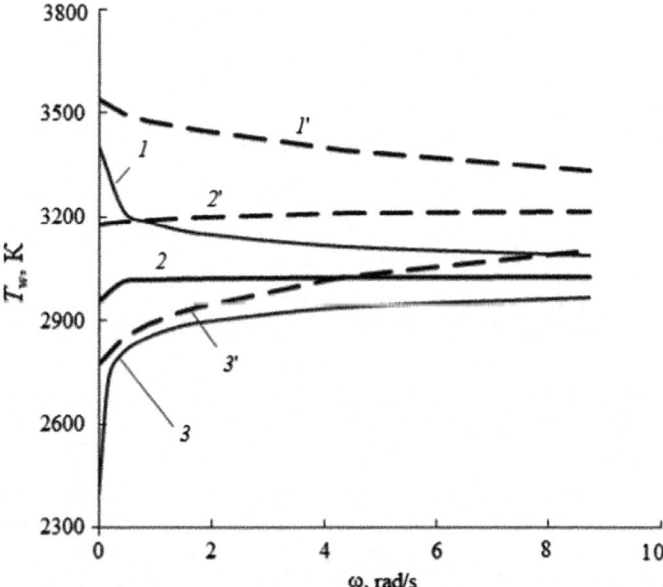

Fig. 3.7 Surface temperature on the conical portion of the body as a function of the rotational speed in the section $\bar{s} = 9.08$. The solid curves correspond to graphite, the dashed curves to CFRP: $1–1'$ are plotted for the maximum surface temperature, $2–2'$ for the average temperature, $3–3'$ for the minimum temperature

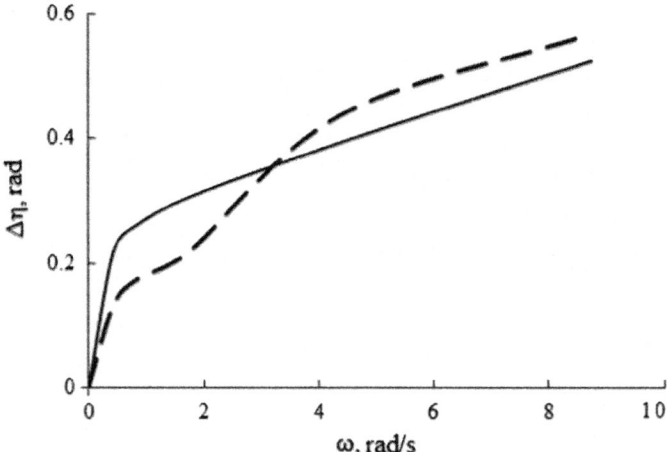

Fig. 3.8 Displacement of the maximum surface temperature in relation to the coordinate $\eta = \pi$ as a function of the rotational speed. The dashed curve corresponds to CFRP, the solid curve to V-1 graphite

(melting point) of the blunted portion is not exceeded throughout the heating area until $t_z = 40$ s.

As we can see in Figs. 3.2 and 3.3, at $\omega \neq 0$ (dashed curves), the surface temperature of the material decreases on the windward side and increases on the leeward side due to heat migration induced by the body rotation.

The temperature of the inside wall of the spherically blunted portion is of practical interest. This temperature reaches $T_{1L_0} = 300$ K at time $t_z = 40$ s. For carbon-fiber-reinforced polymer, the temperature of the inside wall on the conical portion does not exceed $T_{2l} = 306$ K at time t_z.

Figures 3.5 and 3.6 present the distribution of the surface temperature across the circumferential coordinate in the section of the longitudinal coordinate $\bar{s} = 10$ (conical portion) for two materials: porous steel—carbon-fiber-reinforced polymer (Fig. 3.5), permeable steel is V-1 graphite (Fig. 3.6). Dashed curves—_1_, dotted curves—_2_, dash-dotted curves with one dot—_3_, dash-dotted curves with two dots—_4_, and solid curves—_5_ correspond to the rotational speeds of 0, 25, 100, 250, and 500°, respectively.

As seen in these Figures, as the rotation parameter grows, the maximum temperature of the body surface falls, while the minimum temperature increases, i.e., the temperature difference decreases. At the same time, the maximum and minimum body surface temperatures migrate toward the direction of rotation.

Based on the input data from Fig. 3.7 shows the effect of rotation on the differential temperature on the body surface in the section of the longitudinal coordinate $\bar{s} = 10$. The solid and dashed curves correspond to V-l graphite and carbon-fiber-reinforced polymer, respectively. Curves _1–1'_ represent maximum surface temperatures, curves _3–3'_ are minimum surface temperatures, _2–2'_ is average values surface temperatures. The figures show that the differential temperature decreases with the growth of the rotational speed.

A change in the rotational speed has different effects on the differential temperatures at the surface of graphite and carbon-fiber-reinforced polymer: when the rotational speed increases from 0 to 25° (0.436 rad/s), the temperature difference decreases from 1004 to 408 K on graphite and from 765 to 644 K on carbon-fiber-reinforced polymer. The latter change seems to be associated with heat migration induced by the fact that thermal conductivity of graphite is much higher than carbon-fiber-reinforced polymer [15, 33].

With the growth of the rotational speed, it has a weaker effect on the differential temperatures on the body surface as defined by 1, 3 (graphite) and $1'$, $3'$ carbon-fiber-reinforced polymer.

When changing the rotational speed of 0–500° (8.727 rad/s), the average surface temperature $\bar{s} = 10$ in (see the curves 2 and $2'$) is increased due to increased heat transfer to the graphite on 70 K and to the carbon-fiber-reinforced polymer is on 37 K.

Figure 3.8 shows the displacement of the maximum surface temperature in relation to the circumferential coordinate $\eta = \pi$ (windward side) as a function of the rotational speed. This displacement is determined with an accuracy of a node along the circumferential coordinateη and defines the asymmetry of the thermal field at the surface of the thermal-protective material, which results in asymmetry of the flow around the rotating body in relation to the plane of attack angle. Therefore, it gives rise to a negative rolling moment that inhibits rotational motion of the body and a lateral force that deviates the body from a given trajectory. This fact should be taken into account when choosing an acceptable rotational speed. It can be seen that at rotational speeds of more than 250° (4.36 rad/s), the displacement is greater for carbon-fiber-reinforced polymer than for graphite (solid curve).

3.2 Numerical Analysis of Heat and Mass Transfer Characteristics in Radiative and Convective Heating of a Spherically Blunted Cone

At the turn of the twenty-first century, studies of pulse exposure to a highly concentrated energy flow focus on interaction between laser radiation and emerging products of decomposition [5, 125–133]. The studies [125–129] show that exposures of composite materials to pulsed radiation are accompanied by both evaporation and ejection of fragments of a solid-phase material (dispersing of a material). The study [125] demonstrates that the neglect of interaction between ejected products and the energy flow can lead to errors in determining decomposition parameters, as shielding can significantly change spatial and temporal characteristics of the flow. For ebonite and graphite targets [128, 129] are exposed to radiation pulses of moderate intensity $10^8 – 10^{11}$ W/m^2, the screening effect of a laser erosion plasma jet is found to play a significant role. It is shown that condensed-phase particles play a decisive role in screening effects of the laser erosion plasma jet, namely absorption and scattering of laser radiation. In [128], reflection and scattering losses of the laser erosion plasma jet reach the maximum value at a flow density of $\sim 10^{10}$ W/m^2, e.g. 50% in case of ebonite.

Mathematical modeling of heat and mass transfer and combustion processes of composite materials in high-enthalpy flows is discussed in [2, 4, 9, 13]. This paper presents a numerical analysis of carbon fiber reinforced plastic thermochemical decomposition when this material is exposed to laser radiation of moderate intensity $q_* \approx 10^9$ W/m^2. In this regard, the analytical and numerical results of this phenomenon obtained in [134, 135] should be noted. However, no theoretical studies of radiative-convective heating of composite materials were found in the available literature.

In order to describe the carbon fiber reinforced plastic thermochemical decomposition, this paper uses the equations proposed in [4] and the boundary conditions taking into account the "mechanical" ablation of the material from [9]. Unlike the mathematical model [9], the carbon fiber reinforced plastic thermochemical decomposition mechanism cannot be generally described by the model [4]. Furthermore, the conjugate problem of heating a composite body (see Fig. 3.9) within the condensed phase and pyrolysis of the composite material of the shell's conical portion are taken into account.

At relatively small values of laser energy flux density, where an internal problem can be analyzed more or less independently of an external problem to describe surface evaporation kinetics, the so-called thermal model [126] is used. It is based on a solution of the thermal conductivity equation for areas with a moving phase interface and the relevant boundary conditions.

Fig. 3.9 Schematic flow around the body: *1* is porous spherical blunting, *2* is conical part of a body from a composite material

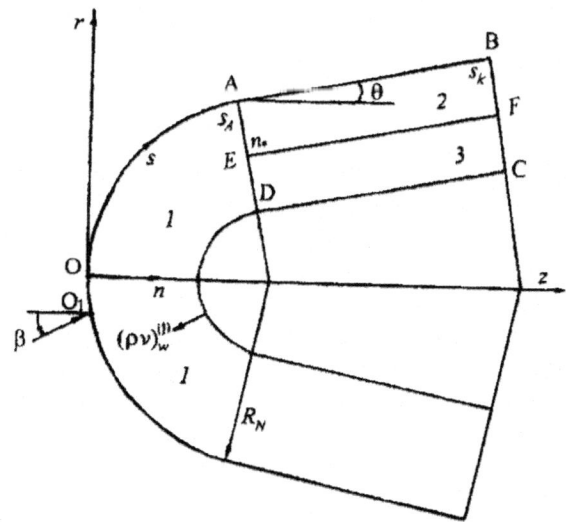

Problem statement. A nine-component model is considered for a conical part of a body. Dominant components of this model are all components that account for five or more percent of the total gas enthalpy $h_w = \sum_{i=1}^{9} h_i c_{iw}$ at least at isolated points of a flow. Therefore, in contrast to [2, 13], heterogeneous chemical reactions presented below are supplemented with a cyanogen formation reaction [5, 136].

We assume that the expression describing attenuation of laser radiation while passing through a laser erosion plasma jet can be expressed as follows (by analogy with the Buger-Lambert law [137])

$$q_r = q_* \exp[-\xi(\rho v)_\Sigma], \tag{3.2.1}$$

where c_{iw} is the mass concentration of the ith component, $i = 1, 2, \ldots, 9$, h is the enthalpy, $(\rho v)_\Sigma$ is the total ablation from the carbon fiber reinforced plastic surface, q_* is the laser energy flux density without blowing out, ξ is the empirical factor $(m^2 \, s)/kg$, which depends on the type of material, optical properties of dispersed-phase particles and a mix of gaseous products of a thermally decomposed composite material. Subscripts $*$ and Σ are attributed to characteristic and total values, respectively, r—to the radiation heat flow, w—to parameters of the body's outer boundary.

The physical meaning of (3.2.1) is evident. The more the total mass rate with which thermochemical decomposition products are injected into the boundary layer, the higher the concentration of condensed particles and polyatomic gaseous decomposition products near a solid surface and the weaker the incident radiant flow.

Furthermore, in order to calculate parameters of the heating process, we should also know the temporal structure of a radiation pulse. In the first approximation, radiation conditions can be described by a step function of time [137]:

$$\begin{cases} W(t) = 0, \, t < t_{2i-1}, \\ W(t) = 1, \, t_{2i-1} \leq t \leq t_{2i}, \, i = 1, 2, \ldots, N, \\ W(t) = 0, \, t > t_{2i}. \end{cases} \tag{3.2.2}$$

The radiation spot is concentrated on the leeward side (see Fig. 3.9) of the conical portion of a composite material in the region $\Delta_* = (\bar{s}_1 \leq \bar{s} \leq \bar{s}_2, \eta_1 \leq \eta \leq \eta_2)$, $\bar{s} = s/R_N$.

Hereinafter, t is the current time, s, η are the components of the natural coordinate system, t_i is the fixed time point, R_N is the radius of a spherically blunted portion, $\bar{s}_1 = 3$, $\bar{s}_2 = 4.771$, $\eta_1 = 180°$, $\eta_2 = 205°$. Next, we took three spikes of laser radiation for numerical calculations: $i = 1, 2, 3$ in (3.2.2) and $t_1 = 10.001$, $t_2 = 10.002$, $t_3 = 10.005$, $t_4 = 10.006$, $t_5 = 10.009$, $t_6 = 10.01$ s.

The condensed-phase process at the conical portion of the body is based on the following physics [2, 4, 9, 13]. Under the influence of a high-temperature flow, the temperature T_2 increases to the resin decomposition temperature. This process is followed by pyrolysis with the formation of carbon residue (coke), which is retained within a reinforcing fiber matrix. At $T_{2w} > 1000$ K, carbon surface is destroyed by reaction with dissociated air components. Pyrolysis gases can be filtered to the interface of media $n = 0$ blown into the boundary layer and, along with products of carbon fiber reinforced plastic oxidation, combustion and sublimation as well as solid-phase particles, and can reduce a radiative and convective heat flow coming to the body.

We make the following assumptions for the problem statement:

(1) the Reynolds number in the incoming hypersonic air flow is sufficiently high ($Re_\infty \gg 1$), and the boundary layer has formed in the vicinity of the body surface;

(2) the air at the outer edge of the boundary layer is in thermochemical equilibrium and composed of five components O, O_2, N, N_2, NO;

(3) transfer processes in the boundary layer are considered on the basis of simplifying assumptions that diffusion coefficients are equal to each other and Lewis number (Le) = 1;

(4) in order to calculate the composition at the interface of gas and condensed phases, we will use the analogy of heat and mass transfer processes [4];

(5) the thermal state of a hemispherical shell in the region 1 (Fig. 3.9) is determined by solving the transient energy conservation equation for the porous spherically blunted portion within the framework of a single-temperature model and the quasi-stationary equation for filtration rate of a coolant gas in pores;

(6) filtered gas in the region 1 is assumed to be uniform and have a molecular mass of air, while density of the gas phase will be determined from the ideal gas equation;

(7) the mass flow rate at a normal n to the surface exposed to a flow is significantly higher than that along s and circumferential η coordinate of this surface;

(8) the following heterogeneous processes take place at the outer surface of the conical portion at $s \geq s_A$ [5, 9, 13, 55, 136]:

$$
\begin{array}{lll}
1)\, C + O_2 \rightarrow CO_2, & 2)\, 2C + O_2 \rightarrow 2CO, & 3)\, C + O \rightarrow CO, \\
4)\, C + CO_2 \rightarrow 2CO, & 5)\, C + N \rightarrow CN, & 6)\, O + O + C \rightarrow O_2 + C, \\
7)\, N + N + C \rightarrow N_2 + C, & 8)\, C \Leftrightarrow C_1, & 9)\, C \Leftrightarrow C_3.
\end{array}
$$

$$(3.2.3)$$

(9) reinforcing filaments of the composite material (carbon fibers) are not melted;

(10) the energy of the laser radiation is absorbed on the surface instantaneously, i.e., the length of the absorption zone in a solid is small in comparison with the depth of heating during the period of interest;

(11) the dispersing condition for particles of a filler (carbon) is assumed to be fulfilled at the maximum density of reinforcing fibers: $\rho_c[x(t), t] = \rho_{c*}$;

(12) the dispersing mechanism is associated with ejection of particles, and velocity of these particles can exceed the filtration rate [9]:

$$v_w^s = -z_1 (\rho v)_{1w} / \rho_{cw}, \ 0 < z_1 < \infty.$$

If serial numbers of components correspond to the following order of enumeration: $O, O_2, N, N_2, CO, CO_2, CN, C_1, C_3$, molar rates of chemical reactions (3.2.3) take the form [9, 13]:

$$U_1 = \frac{k_1 c_{2w} \rho_w^{(2)}}{m_2} \exp\left(-\frac{E_1}{RT_{2w}}\right), \qquad U_2 = \frac{k_2 c_{2w} \rho_w^{(2)}}{m_2} \exp\left(-\frac{E_2}{RT_{2w}}\right),$$

$$U_3 = \frac{k_3 c_{1w} \rho_w^{(2)}}{m_1} \exp\left(-\frac{E_3}{RT_{2w}}\right), \qquad U_4 = \frac{k_4 c_{6w} \rho_w^{(2)}}{m_6} \exp\left(-\frac{E_4}{RT_{2w}}\right)$$

$$U_5 = \frac{k_5 c_{3w} \rho_w^{(2)}}{m_3}, \ U_6 = \frac{k_6 c_{1w} \rho_w^{(2)}}{m_1}, \ U_7 = \frac{k_7 c_{3w} \rho_w^{(2)}}{m_3}, \tag{3.2.4}$$

$$U_i = \frac{A_{ci}(P_{ci}^* - P_{ci})}{(2\pi RT_{2w} m_i)^{0.5}}, \qquad P_{ci}^* = 10^5 \exp\left(D_i - \frac{E_i}{T_w}\right),$$

$$m_w^{-1} = \sum_{\alpha=1}^{9} \frac{c_{\alpha w}}{m_\alpha}, \qquad P_{ci} = \frac{P_e c_{iw} m_w}{m_i}, \ i = 8, 9.$$

By using (3.2.4), we will find the mass rates of generation (disappearance) of components as a result of heterogeneous reactions:

$$R_1 = -m_1(U_3 + U_6), \ R_2 = -m_2(U_1 + U_2 - U_6/2), \ R_3 = -m_3(U_5 + U_7),$$

$$R_4 = m_4 U_7/2, \ R_5 = m_5(2U_2 + U_3 + 2U_4), \ R_6 = m_6(U_1 - U_4),$$

$$R_7 = m_7 U_5, \ R_i = A_{ci} m_i (P_{ci}^* - P_{ci})(2\pi RT_w m_i)^{0.5} \ i = 8, 9.$$

Mass ablation rates can be expressed as [13]:

$$(\rho v)_{2w} = \varphi_2 \rho_w^{(2)} \left[\left(\frac{m_6}{m_2} - 1\right) c_{2w} B_1 + \left(2\frac{m_5}{m_2} - 1\right) c_{2w} B_2 \right.$$

$$\left. + \left(\frac{m_5}{m_1} - 1\right) c_{1w} B_3 + \left(2\frac{m_5}{m_6} - 1\right) c_{6w} B_4 + \left(\frac{m_7}{m_3} - 1\right) k_5 c_{3w} \right],$$

$$(\rho v)_{3w} = \varphi_2 \sum_{i=8}^{9} \frac{m_i A_{ci}(P_{ci}^* - P_{ci})}{(2\pi RT_{2w} m_i)^{0.5}}, \quad B_i = k_i \exp(-E_i/RT_{2w}), i = \overline{1, 4}, \tag{3.2.5}$$

$$(\rho v)_w^{(2)} = \sum_{i=1}^{3} (\rho v)_{iw}, \quad (\rho v)_\Sigma = (\rho v)_w^{(2)} + z_1 (\rho v)_{1w}.$$

The seventh and ninth assumptions are justified in [2] and [136], respectively. C is the designation of solid carbon, which is contained in the thermal-protective coating. There are five components in the boundary layer: O, O_2, N, N_2, NO, which are involved in three equilibrium chemical reactions: $O_2 \leftrightarrow 2O, N_2 \leftrightarrow 2N$, $NO \leftrightarrow N + O$. There are five components at the interface of condensed and gas phases: CO, CO_2, CN, C_1, C_3, which are generated by seven heterogeneous reactions of combustion and sublimation (3.2.3). Two catalytic recombination reactions of the components O_2, N_2 are taken into account here [55]. We neglect argon in the boundary layer and the sublimation component C_2 at the surface of the cone, since, according to [5, 136], their values in the considered temperature range are less than 1 and 5%, respectively.

Subject to the assumptions (5)–(7) for the porous spherical shell $0 < s < s_A$, in case of one-dimensional filtration of the gas injected toward the normal to the surface in the coordinate system tied to the symmetry axis of the body, we have [2, 9, 13]:

$$\frac{\partial[(\rho v)^{(1)} \varphi_1 r_1 F_1]}{\partial n} = 0, \tag{3.2.6}$$

$$(\rho c_p)_1 (1 - \varphi_1) \frac{\partial T_1}{\partial t} = \frac{1}{r_1 F_1} \left\{ \frac{\partial}{\partial n} \left[r_1 F_1 \lambda_1 (1 - \varphi_1) \frac{\partial T_1}{\partial n} \right] \right.$$
$$\left. + \frac{\partial}{\partial s} \left[\frac{r_1 \lambda_1}{F_1} (1 - \varphi_1) \frac{\partial T_1}{\partial s} \right] + \frac{\partial}{\partial \eta} \left[\frac{F_1 \lambda_1}{r_1} (1 - \varphi_1) \frac{\partial T_1}{\partial \eta} \right] \right\} + c_{pg}^{(1)} (\rho v)_w^{(1)} \frac{r_{1w}}{r_1 F_1} \frac{\partial T_1}{\partial n}, \tag{3.2.7}$$

$$A \mu_1 v^{(1)} + B \varphi_1 \rho^{(1)} v^{(1)} \left| v^{(1)} \right| = -\frac{\partial P}{\partial n}, \tag{3.2.8}$$

$$P = \frac{\rho^{(1)} R T_1}{M}, \quad F_1 = \frac{R_N - n}{R_N}, \tag{3.2.9}$$

$$r_1 = (R_N - n) \sin(\bar{s}), \quad \mu_1 \sim \sqrt{T_1}, \quad \lambda_1 \sim \sqrt{T_1}, \quad \varphi_1 = \text{const}.$$

For the conical portion of the body $(s_A < s < s_k)$, the mass and energy conservation equations in the movable coordinate system can be written on the basis of the mathematical models proposed in region 2 and 3 on a Fig. 3.9 recording [4, 9]:

$$\rho_c c_{p2} \left(\frac{\partial T_2}{\partial t} - \psi \frac{\partial T_2}{\partial n} \right) + c_{pg}^{(2)} G \frac{\partial T_2}{\partial n} = \frac{\partial}{\partial n} \left(\lambda_2 \frac{\partial T_2}{\partial n} \right)$$
$$+ \frac{\partial}{\partial s} \left(\lambda_2 \frac{\partial T_2}{\partial s} \right) + \frac{1}{r_2^2} \frac{\partial}{\partial \eta} \left(\lambda_2 \frac{\partial T_2}{\partial \eta} \right) - Q_c \frac{d \rho_c}{dt}, \tag{3.2.10}$$

$$\frac{d \rho_c}{dt} = \left(\frac{\partial \rho_c}{\partial t} - \psi \frac{\partial \rho_c}{\partial n} \right)$$
$$= \begin{cases} -k_c \rho_{c0} \left(\frac{\rho_c - \rho_{c*}}{\rho_{c0}} \right) \exp\left(-\frac{E_c}{R T_2} \right), & \rho_c > \rho_{c*}, \\ 0, & \rho_c \leq \rho_{c*}, \end{cases} \tag{3.2.11}$$

$$c_{pg}^{(i)} = b_1 + b_2 T_i, \ i = 1, 2, \ r_2 = (R_N - n) \cos \theta + (s - s_A) \sin \theta,$$

$$G = \int\limits_0^l \frac{d\rho_c}{dt} dn, \ \psi = \left[\sum_{i=2}^3 (\rho v)_{iw} + z_1 (\rho v)_{1w} \right] / \rho_{cw}, \ (\rho v)_{1w} = G_w,$$

and for graphite V-1 at $i = 2$ in region 2 and substrate from asbestos cement in region 3 at $i = 3$ we obtain

$$\rho_i c_{pi} \left(\frac{\partial T_i}{\partial t} - \psi \frac{\partial T_i}{\partial n} \right) = \frac{\partial}{\partial n} \left(\lambda_i \frac{\partial T_i}{\partial n} \right)$$
$$+ \frac{\partial}{\partial s} \left(\lambda_i \frac{\partial T_i}{\partial s} \right) + \frac{1}{r_2^2} \frac{\partial}{\partial \eta} \left(\lambda_i \frac{\partial T_i}{\partial \eta} \right), \ i = 2, 3. \tag{3.2.12}$$

In order to determine the temperature of dispersed-phase particles, [9] we use $T_{2w}^s = T_2(t - t_*)$, where t_* is the time to reach the maximum density ρ_{c*}. This formula corresponds to the case where the detachment time t_* was not sufficient for a particle to exchange energy with the ambient gas medium, and the particle maintained the same temperature as it had at the beginning of the detachment.

The system of Eqs. (3.2.6)–(3.2.8), (3.2.10)–(3.2.12) needs to be solved with regard to the initial and boundary conditions:

Initial conditions:

$$T_i|_{t=0} = T_0, \ i = 1, 2, 3, \ \rho_c|_{t=0} = \rho_{c0}. \tag{3.2.13}$$

Boundary conditions.

At the external shell surface exposed to a flow ($n = 0$), the conditions [2, 13] take place at $0 \le \eta < 2\pi$

$$q_{1w} - (1 - \varphi_1)\varepsilon_1 \sigma T_{1w}^4 - -\lambda_1(1 - \varphi_1)(\partial T_1 / \partial n)|_w, \tag{3.2.14}$$

$$Q_\Sigma = -\lambda_2(\partial T_2 / \partial n)|_w,$$
$$Q_\Sigma = Q_w, \ t < t_{2i-1}, \ t > t_{2i}, \ i = 1, 2, 3, \tag{3.2.15}$$
$$Q_\Sigma = Q_w + W q_r, \ t_{2i-1} \le t \le t_{2i},$$

$$Q_w = q_{2w} - (h_w - h_c) \sum_{i=2}^3 (\rho v)_{iw} - z_1 (\rho v)_{1w} \left(h_w^s - h_c \right)$$
$$- (\rho v)_{1w} \left(h_w - h_g \right) - \varphi_2 \varepsilon_2 \sigma T_{2w}^4, \tag{3.2.16}$$

$$q_{1w} = \alpha_1 (h_{e0} - H_w), \ q_{2w} = \alpha_2 (h_{e0} - h_w), \ h_w^s = c_{p2} T_{2w}^s,$$
$$h_g = \int\limits_0^{T_2} c_{pg}^{(2)} dT_2, \ H_w = T_{1w} b_1 + b_2 T_{1w}^2 / 2. \tag{3.2.17}$$

The internal surface of the hemisphere DE and the conical portion of the body DC is subject to the following relations [2]:

$$\lambda_1(1 - \varphi_1)\partial T_1/\partial n|_{n=L_0} = \delta(T_{1,L_0} - T_0), \; 0 \leq s < s_A, \quad (3.2.18)$$

$$\rho_c|_{n=l} = \rho_{c0}, \quad \lambda_3(\partial T_3/\partial n)|_{n=l} = 0, \; s_A \leq s \leq s_k. \quad (3.2.19)$$

The perfect contact conditions are used at the sphere-cone interface ring AD at $s = s_A$ and on the line EF at $n = n_*$:

$$\frac{\lambda_1(1 - \varphi_1)}{F_1}\frac{\partial T_1}{\partial s}\bigg|_{s=s_A-0} = \lambda_i\frac{\partial T_i}{\partial s}\bigg|_{s=s_A+0}, \quad (3.2.20)$$

$$T_1|_{s=s_A-0} = T_i|_{s=s_A+0}, \; i = 2,3$$

$$\lambda_2\frac{\partial T_2}{\partial n}\bigg|_{n=n_*-0} = \lambda_3\frac{\partial T_3}{\partial n}\bigg|_{n=n_*+0}, \quad T_2|_{n=n_*-0} = T_3|_{n=n_*+0}, \quad (3.2.21)$$

In the conical portion BC: $s = s_k$, an adiabatic condition is used:

$$(\partial T_i/\partial s)|_{s=s_k} = 0, \; i = 2,3. \quad (3.2.22)$$

Pressures in pores and the environment are equal to each other on external and internal surfaces of the spherically blunted area.

$$P_w|_{n=0} = P_e(s,\eta), \; P|_{n=L_0} = P_{L_0}. \quad (3.2.23)$$

When the flow does not have a plane of symmetry, periodicity conditions are fulfilled:

$$T_i(t,n,s,\eta) = T_i(t,n,s,\eta+2\pi), \; i = 1,2,3 \quad (3.2.24)$$

where

$$l = L_0 - x(t), \; x(t) = \int_0^t \psi\, d\tau.$$

Hereinafter, A and B are the viscosity and inertial coefficients in the nonlinear Darcy law (3.2.8); r_{1w}, r_i, $i = 1, 2$, F_1 are the Lame coefficients; $(\rho v)_w^{(1)}$ is the flow rate of coolant gas from the spherically blunted surface; G is the mass flow rate of gaseous carbon fiber reinforced plastic decomposition products; $(\rho v)_w^{(2)}$ is the total ablation from the carbon surface of the conical portion without regard to dispersion; ρ is the density; $\rho_c\psi$ is the mass velocity of the destructed surface; $Q_c d\rho_c/dt$ is the heat absorbed as a result of carbon fiber reinforced plastic pyrolysis reaction; φ_1 is the porosity of the spherically blunted surface; M is the molecular mass; V is the velocity; φ_2 is the percentage of the frame of the conical portion; M_∞ is the Mach number; H_∞ is the flight altitude of the body; L_0 is the initial thickness of the shell; θ is the taper angle; β is the angle of attack; n is the normal to the surface toward the inside of the shell; ψ is the linear velocity of the destructed surface; $x(t)$ is the interface between gas and condensed phases (burn-up depth); $E_i, k_i, i = 1, ..., 4$ are, respectively, the activation energy and the pre-exponential factor of the ith heterogeneous reaction on the shell of the conical portion; k_c, E_c and Q_c are, respectively, the pre-exponential factor, the activation energy and the thermal effect of the pyrolysis reaction.

The subscripts e and e0 correspond to the values at the outer edge of the boundary layer and in the outer edge at the stagnation point of the body, respectively; subscripts 1, 2, 3—to the characteristics in the sphere, and in the cone in regions 2 and 3, either to carbon fiber reinforced plastic or B-1 graphite or in the region 2—graphite and in the region 3—asbestos cement; g—to the gas phase on spherical and conical portions of the body,—to the values of the incoming gas flow at infinity; lam, tur, 0—to the characteristics of laminar, turbulent transfer, and the initial conditions, respectively; L—to the internal shell of the spherical portion; k—to the peripheral region of the shell. The superscripts (1) and (2) correspond to the characteristics of the gas phase on the sphere and cone of the body, respectively; the overline—to non-dimensional parameters; z—to the end time of thermal exposure; ef—to the effective value; c—to carbon fiber reinforced plastic; *—to the characteristic quantity, m—to the maximum value.

Let us write the balance relationships for mass concentrations of components (c_{iw}) using Fick's law for diffusion flows, as well the analogy of heat and mass transfer processes [4, 9]:

$$J_{iw} + (\rho v)_w^{(2)} c_{iw} = R_{iw}, \quad i = \overline{1,9}, \quad J_{iw} = \beta_i(c_{iw} - c_{ie}), \quad \beta_i = \alpha/c_p,$$

where α/c_p and β_i are the heat and mass transfer coefficients, respectively. Decomposition products are considered to weakly dilute the air mixture in the boundary layer. This assumption makes it possible to use the setting accepted above for the equations in the boundary layer.

In order to define a heat flow from the gas phase q_w, we can use the formulas presented in [138] for a spatial case under laminar and turbulent flow conditions in the boundary layer. For attenuating the heat flow with injected coolant gas that has the same composition as the incoming air flow, we will use the formulas from [11]. Under laminar flow conditions in the boundary layer [2, 4, 138], we obtain the following results for the porous spherical portion in the coordinate system tied to the stagnation point:

$$q_{1w} = \alpha_1 \exp[-0.6(\rho v)_w^{(1)}/\alpha_1](h_{e0} - H_w), \quad \alpha_{lam} = 1.05 V_\infty^{1.08}$$
$$\times (\rho_\infty/R_N)^{0.5} \times [0.55 + 0.45\cos(2\tilde{s})],$$

$$\tilde{s} = \arccos(\cos\bar{s}\cos\beta + \sin\bar{s}\sin\beta\cos\eta),$$
$$h_{e0} = V_\infty^2/2 + h_\infty, \ \alpha_1 = \alpha_{lam}, \ 0_1 \le \tilde{s} \le \tilde{s}_*. \tag{3.2.25}$$

For turbulent flow conditions in the boundary layer [2, 138], we have:

$$q_{1w} = \alpha_1 \, \exp[-0.37(\rho v)_w^{(1)}/\alpha_1] \quad (h_{e0} - H_w),$$

$$\alpha_{tur} = \frac{16.4V_\infty^{1.25}\rho_\infty^{0.8}}{R_N^{0.2}(1 + H_w/h_{e0})^{2/3}} \times (3.75\sin\tilde{s} - 3.5\sin^2\tilde{s}),$$

$$\alpha_1 = \alpha_{tur}, \ \tilde{s}_* < \tilde{s} < \tilde{s}_A, \tag{3.2.26}$$

where $\tilde{s}_*$ are coordinates of the instability point in the coordinate system referenced to the stagnation point.

In order to estimate the effect of injection on the heat flow in the curtain zone, we can use the findings from [139] and the formulas from [96] based on the processed high-accuracy numerical calculations for the boundary layer and the viscous shock layer

$$q_{2w} = \alpha_2 \left(1 - k_1 b^{k_2}\right)(h_{e0} - h_w),$$
$$\alpha_2 = \alpha_w \, \exp\left[-0.37(\rho v)_w^{(2)}/\alpha_w\right], \tag{3.2.27}$$

$$\alpha_w = \frac{16.4V_\infty^{1.25}\rho_\infty^{0.8}}{R_N^{0.2}(1 + h_w/h_{e0})^{2/3}}, \ \frac{2.2\bar{p}(u_e/v_m)}{\varsigma^{0.4}\bar{r}_2^{0.2}}, \ \bar{r}_2 = \cos\theta + (\tilde{s} - \tilde{s}_A)\sin\theta,$$

$$\bar{p} = P_e/P_{e0}, \ u_e/v_m = (1 - \bar{p}^\chi)^{0.5}, \ \varsigma = (\gamma_{ef} - 1 + 2/M_\infty^2)/(\gamma_{ef} + 1),$$

$$\tilde{s}_A \le \tilde{s} \le \tilde{s}_B, \ \chi = (\gamma_{ef} - 1)/\gamma_{ef}.$$

According to the law defining the flow rate of the coolant gas:

$$(\rho v)_w(\tilde{s}) = (\rho v)_w(0_1)\left(1 + a\sin^2\tilde{s}\right),$$

we have:

$$b = \frac{2(\rho v)_w(0_1)\{1 - \cos\tilde{s}_A + a[2/3 - \cos\tilde{s}_A + 1/3\cos^3\tilde{s}_A]\}}{\alpha_w(\tilde{s} - \tilde{s}_A)[2\cos\theta + (\tilde{s} - \tilde{s}_A)\sin\theta]}, \tag{3.2.28}$$

$$\cos\tilde{s}_A = \cos\bar{s}_A\cos\beta + \sin\bar{s}_A\sin\beta\cos\eta, \ \bar{s}_A = \pi/2 - \theta.$$

Calculation procedure, tests and initial data. The boundary problems (3.2.6)–(3.2.8), (3.2.10)–(3.2.12), (3.2.13)–(3.2.15), (3.2.18)–(3.2.24) were solved numerically by locally one-dimensional splitting method [82]. We used the implicit, totally stable monotonic difference scheme with total approximation error $O(\tau + H_\eta^2 + H_n^2 + H_s^2)$, where H_n is the spatial step along the coordinate n, H_s is the spatial step along the coordinate s, H_η is the spatial step along the coordinate η, τ is the time step. A sequence of spatially condensing grids was used for testing the numerical calculation algorithm. $h_1 = h_n = 5\times10^{-4}$ m, $h_2 = h_{s1} = 8.75\times10^{-2}$ (at the sphere), $h_3 = h_{s2} = 8.55\times10^{-2}$ (at the cone), $h_4 = h_\eta = 0.087$, with $H_{1,i} = 2 \times h_i$, $H_{2,i} = h_i$, $H_{3,i} = h_i/2$, $H_{4,i} = h_i/4$, $i = 1, 2, 3, 4$. The temperature of the frame was recorded across the depth of the body at different time points. In all cases, the problem was solved with the variable time step chosen on the assumption that the prescribed accuracy was equal for all spatial steps. The difference in the relative temperature error decreased and, by $t = t_z$ reached $\Delta_1 = 9.3\%$, $\Delta_2 = 5.2\%$, $\Delta_3 = 2.4\%$. The calculation results presented below were obtained for spatial steps $H_{3,i} = h_i/2$, $i = 1, 2, 3, 4$.

The quasi-stationary equation of continuity (3.2.6) $(\rho v)_w^{(1)} r_{1w}/(F_1 r_1) = -\rho^{(1)}\varphi_1 v^{(1)}$ (the negative sign is explained by the fact that the normal coordinate n is directed deep into the body (see Fig. 3.9), and the coolant flows in the opposite direction) can be integrated with the first equation (3.2.9), the nonlinear Darcy law (3.2.8) and the boundary conditions (3.2.23) in order to find the gas flow rate and the pressure in the region 1 [2]:

$$(\rho v)_w^{(1)}(s, \eta, t) = \frac{\left[2B(P_{L_0}^2 - P_w^2)\varphi_1 MDL_0/R + E_{L_0}^2\right]^{0.5} - E_{L_0}}{2BDL_0}, \qquad (3.2.29)$$

$$P(n, s, \eta, t) = \left\{P_w^2 + 2R(\rho v)_w^{(1)}\left[B(\rho v)_w^{(1)}D + E\right]/M\,\varphi_1\right\}^{0.5},$$

where $D(n, s, \eta, t) = \int_0^n T_1 \left(\frac{r_{1w}}{r_1 F_1}\right)^2 dy$, $E(n, s, \eta, t) = A\int_0^n \mu T_1 \frac{r_{1w}}{r_1 F_1} dy$.

The pressure on the inside "cold" surface of the spherical shell ($n = L_0$) is given by

$$P_{L_0} = kP_{e0},$$

where k is a certain constant. This ensured a necessary flow rate of the coolant (in particular, the melting temperature of the porous metal frame was not reached [4]) in the thermal exposure area from $t = 0$ to $t = t_z$ ($t_z = 12$ s corresponds to the end time of thermal exposure).

The flow of chemical-equilibrium air at an angle of attack $\beta = 5°$ around the spherically blunted cone with a one-half angle $\theta = 10°$ was calculated for the following conditions: $T_{e0} = 6739$ K, $P_{e0} = 13.59 \times 10^5$ N/m^2, $V_\infty = 5.5 \times 10^3$ m/s, $H_\infty = 2.4\times10^4$ m, $R_N = 0.2$ m, $L_0 = 0.01$ m. The kinetic constants (3.2.4) of the reactions (3.2.3) were taken from [9, 140]. The enthalpy of graphite h_c was calculated by formula [58]. The effective adiabatic exponent γ_{ef} in the latter formula (3.2.27) and dimensionless pressure $\bar{p} = P_e/P_{e0}$ were determined according to [115]. Enthalpies h_i, $i = \overline{1,9}$ and heat capacities of the components involved in the chemical reactions were derived from [141].

Thermal characteristics of the porous blunted portion correspond to porous copper: $\varepsilon_1 = 0.8$, $\lambda_1 = 2.92 + 4.5 \times 10^{-3} \times T_1$ W/(m K), $\rho_1 c_{p1} = (1252 + 0.544 \times T_1) \times 10^3$ J/(K m^3), $A = 2.3 \times 10^{11}$ 1/m^2, $B = 5.7 \times 10^5$ 1/m [30]. Thermal characteristics of the conical portion correspond to carbon fiber reinforced plastic [2], or V-1 graphite [124] and asbestos cement [83].

$$\lambda_2 = 3.2 \times 10^{-4} \rho_c, \ 293 \leq T_2 600 \, \text{K},$$

$$\lambda_2 = 3.2 \times 10^{-4} \rho_c + 2.1 \times 10^{-3}(T_2 - 600), \ 600 \leq T_2 \leq 1400 \, \text{K},$$

$$c_{p2} = 950 + 0.7364(T_2 - 293), \ T_2 \leq 1400 \, \text{K},$$

$$\lambda_2 = 1.7 + 0.021 \times T_2^{0.5}, \ c_{p2} = 1236 + 13.4 \cdot T_2^{0.5}, \ T_2 > 1400 \, \text{K}.$$

The results presented below were obtained at $\varphi_1 = 0.34$, $\varphi_2 = 0.9$, $T_0 = 293$ K, $M_\infty = 18$, $\rho_\infty = 4.75 \times 10^{-3}$ (c$^2\cdot$kg)/m^4, $\delta = 100$ W/(m^2 K), $M = 29$ kg/kmol, $\sigma = 5.67 \times 10^{-8}$ W/(m^2 K^4), $\varepsilon_2 = 0.9$, $\rho_{c0} = 1400$ kg/m^3, $\rho_{c*} = 1300$ kg/m^3, $q_* = 10^9$ W/m^2, $R = 8.314$ J/(mole K), $k_c = 3.15 \times 10^6$ s^{-1}, $E_c = 8.38 \times 10^4$ J/mole, $Q_c = 1.26 \times 10^6$ J/kg, $\xi = 1$ (m^2 s)/kg, $m_1 = 16$ kg/kmol, $m_2 = 32$ kg/kmol, $m_3 = 14$ kg/kmol, $m_4 = 28$ kg/kmol, $m_5 = 28$ kg/kmol, $m_6 = 44$ kg/kmol, $m_7 = 26$ kg/kmol, $m_8 = 12$ kg/kmol, $m_9 = 36$ kg/kmol, $\rho_3 = 1800$ kg/m^3, $c_{p3} = 837$ J/(kg K), $\lambda_3 = 0.349$ W/(m K) $z_1 = 1$, $b_1 = 965.5$, $b_2 = 0.147$, $a = 1$, $k = 1.3$, $k_1 = 0.285$, $k_2 = 0.165$.

Discussion of the numerical solution. Figure 3.10a presents the convective heat flow q_{2w} from the gas phase (1) and the surface enthalpy h_w (2) as a function of time. Figure 3.10b shows profiles of CFRP surface temperature (1) and the total heat flow to the condensed phase (2) in the fixed section ($\bar{s}_1 = 3$, $\eta_1 = 180°$) of the blunted cone in the curtain zone as a function of time. The solid curves correspond to the characteristics of carbon fiber reinforced plastic, and the dashed curves correspond to the parameters of graphite V-1 without a substrate.

Dependences for the mass concentration of components (solid curves) and ablation (dashed curves) on the surface of carbon fiber are shown in Fig. 3.11a. Similar dependences for the concentrations of components and ablation are given on the graphite surface in Fig. 3.11b. It is seen that when the high-enthalpy flow is applied, the surface temperature T_{2w} increases, carbon monoxide c_{5w} [13] appears, and ablation occurs $(\rho v)_\Sigma$. When the first radiation pulse is turned on $t > t_1$, the surface temperature increases (Fig. 3.11b), and the concentration of cyanide c_{7w} begins to increase (see Fig. 3.11) and a subliming component c_{9w} appears (see Fig. 3.11).

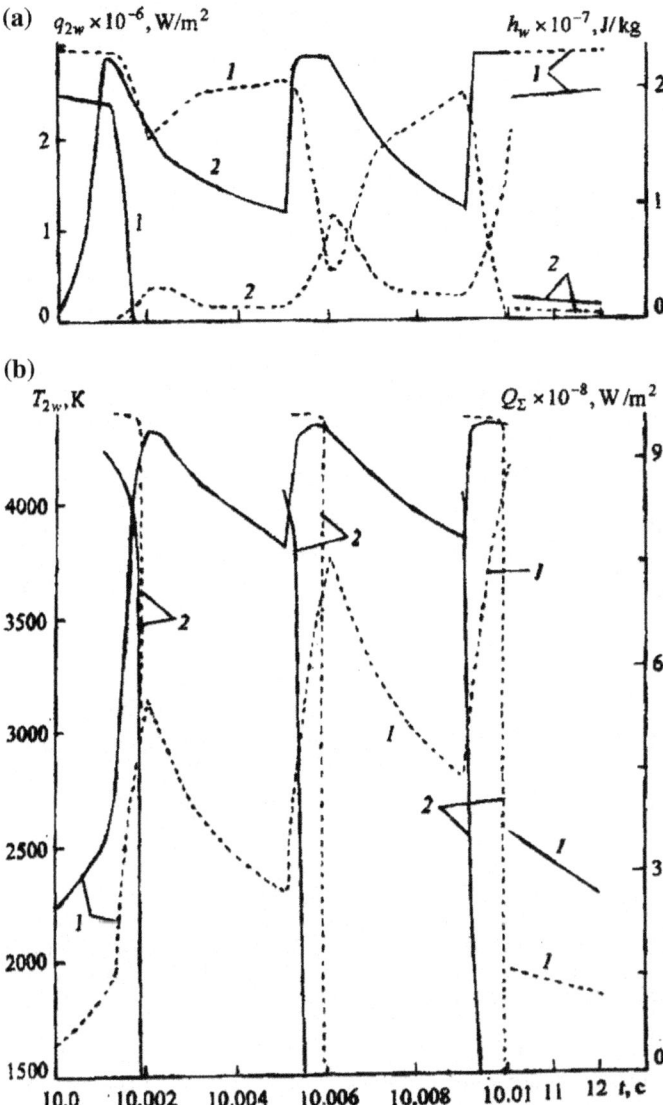

Fig. 3.10 Dependence of the convective heat flux from the gas phase (*1*) and the enthalpy of the surface (*2*) in a fixed section of the conical part of the body versus time (**a**). Dependence of the temperature of the surface (*1*) and the total heat flux in the condensed phase (*2*) in a fixed section on time (**b**). The solid curves correspond to carbon plastic, dashed is graphite V-1 without a substrate

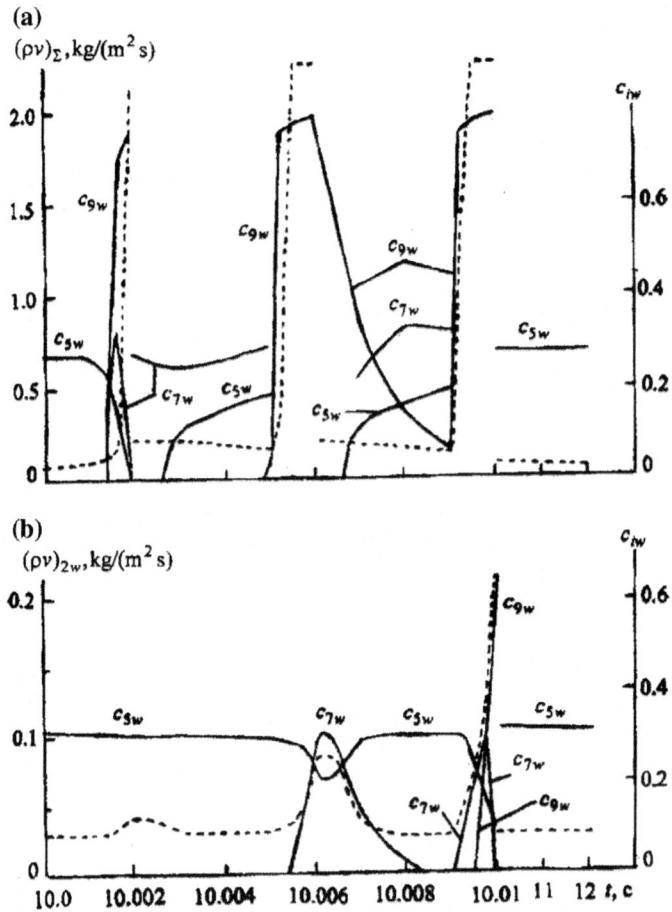

Fig. 3.11 Dependence of the mass concentration of components (solid line) and the total mass ash (dashed curve) in the fixed section from a time. Figure 3.11, **a** corresponds to the characteristics of CFRP, and Fig. 3.11, **b**—parameters of graphite V-1 without a substrate

At the same time, the value of the convective flow q_{2w} decreases (see line _1_ in Fig. 3.10a), since T_{2w} and the enthalpy h_w increases (line _2_ in Fig. 3.10a) due to the growth of components c_{7w} and c_{9w} (see Fig. 3.11). In addition, q_{2w} is weakening according to the formula (3.2.27), since products of carbon fiber reinforced plastic thermochemical.

Note that the radiant flux q_r (3.2.1) and the total heat flux Q_Σ (3.2.15) into the condensed phase (Fig. 3.10b) also decrease. Then, over time, the value $(\rho v)_\Sigma$ falls for carbon fiber reinforced plastic due to the screening effect of the coked layer [2, 9], the thickness of which increases with the progress of the carbon fiber reinforced plastic [13]. After termination of radiant heating ($t > t_2$), T_{2w} decreases, since the body is exposed to a relatively "cold" (as compared to the surface temperature) convective heat flow from the gas phase (Fig. 3.10a). This in turn leads to a slowing down of the pyrolysis of carbon fiber and to a decrease $(\rho v)_\Sigma$ (see the dashed curve in Fig. 3.11).

After the exposure to the next laser spikes t_i, $i = 2, 3$, the distribution pattern of the heat and mass transfer characteristics listed above is repeated. However, when the body is preheated, the surface temperature in the action zone of the second laser radiation pulse is higher than that in the first spike. Therefore, the magnitude $(\rho v)_\Sigma$ of the second and third pulsed loading in the radiant warm-up section is determined by sublimation of the carbon surface of the body (see solid curves c_{9w} on Fig. 3.11 for $t_3 \leq t \leq t_6$). The latter factor leads to attenuation of the radiation pulse according to the formula (3.2.1) and is associated with the fact that the radiant component is shielded with both products of heterogeneous chemical reactions and dispersed-phase particles. As a result, Q_Σ Fig. (3.11b) decreases due to the heat removed during repeated radiation and sublimation of the carbon fiber reinforced plastic surface. In the absence of radiation heating: $t > t_6$ the material in the conical portion $\Delta_* = (\bar{s}_1 \leq \bar{s} \leq \bar{s}_2, \eta_1 \leq \eta \leq \eta_2))$ of the body is heated with the total heat flow Q_w from (3.2.16), the value of which is two orders of magnitude lower than Q_Σ for $t \leq t_6$ and, therefore, not shown (Fig. 3.10b) at $t > t_6$.

Since Q_w falls, the carbon fiber reinforced plastic heating process slows down, while the surface temperature decreases until the end of thermal exposure. As expected, $(\rho v)_\Sigma$ in this period is largely determined by components $(\rho v)_{1w}$, $(\rho v)_{2w}$ from (3.2.5), (3.2.12), and while the total heat flow Q_w is determined by the heating component attributed to dispersing and pyrolysis of the composite material.

Figures 3.12, 3.13 and 3.14 show the surface temperature T_{1w}, T_{2w}(the solid curves correspond to the characteristics of CFRP, the dashed curves correspond to graphite), ablation from the conical portion $(\rho v)_\Sigma$ (3.2.5), the spherically blunted portion (solid lines 5) $(\rho v)_w^{(1)}$ (3.2.29), and the convective heat flow q_{1w}, q_{2w} in the symmetry plane of the flow (3.2.25), (3.2.26), (3.2.27) and (3.2.28) ($\eta = 0$–180°) in windward and leeward sides as a function of the longitudinal coordinate $\bar{s}$. Lines 1–4 in Figs. 3.12, 3.13, and 3.14 correspond to times t: 1–5, 2–10.002, 3–10.01, 4–12 s. At the same time points on Figs. 3.15 and 3.16 present distributions of the surface temperature T_{2w} and total ablation $(\rho v)_\Sigma$ from the conical portion as a function of the circumferential coordinate η for $\bar{s}_1 = 3$. As it is seen from Figs. 3.12 and 3.15, the maximum surface temperature of the body is recorded on the leeward side of the cone Δ_*, in the spot exposed to laser radiation. In the same region Δ_* on the leeward side of the cone, there is maximum ablation from the surface (see Figs. 3.13 and 3.16), which can shield (attenuate) the radiant heat flow according to the formula (3.2.1).

According to analyzed heat and mass transfer characteristics of a steel spherical shell (Figs. 3.12, 3.13, and 3.14), the prescribed injection of coolant gas (5 curve line in Fig. 3.13) ensures that the critical temperature of the blunted portion will not exceed $T_* = 1600$ K [8] until $t = t_z$ due to attenuation of the heat flow as per (3.2.25) and (3.2.26). Moreover, the heat is absorbed when gas is filtered into pores

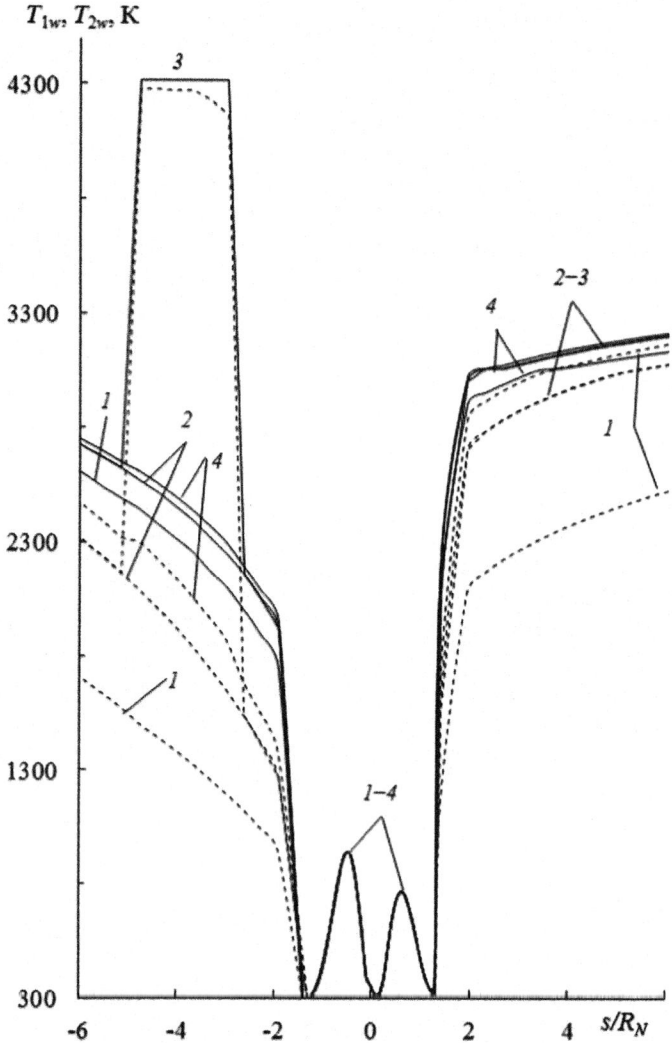

Fig. 3.12 Distribution of surface temperature in the plane of symmetry of the flow ($\eta = 0$–$180°$) on the windward and leeward side of the body on the longitudinal coordinate. Curves correspond *1–6* moments of time *t*: *1*–5, *2*–10.002, *3*–10.01, *4*–12 s. The solid curves correspond to the characteristics of CFRP, and the dashed curves correspond to the parameters of graphite V-1 without a substrate

of the permeable spherically blunted portion. The maximum surface temperature on the porous sphere corresponds to the maximum heat flow from the gas phase.

For practical purposes, the value of the internal wall temperature of the spherical blunting, which for the permeable steel was $T_{1L_0} = 293$ K at $L_0 = 0.01$ m at the moment $t = t_z$. On the conic part of the body in cross section: $\bar{s}_1 = 3$, $\eta = 180°$ the

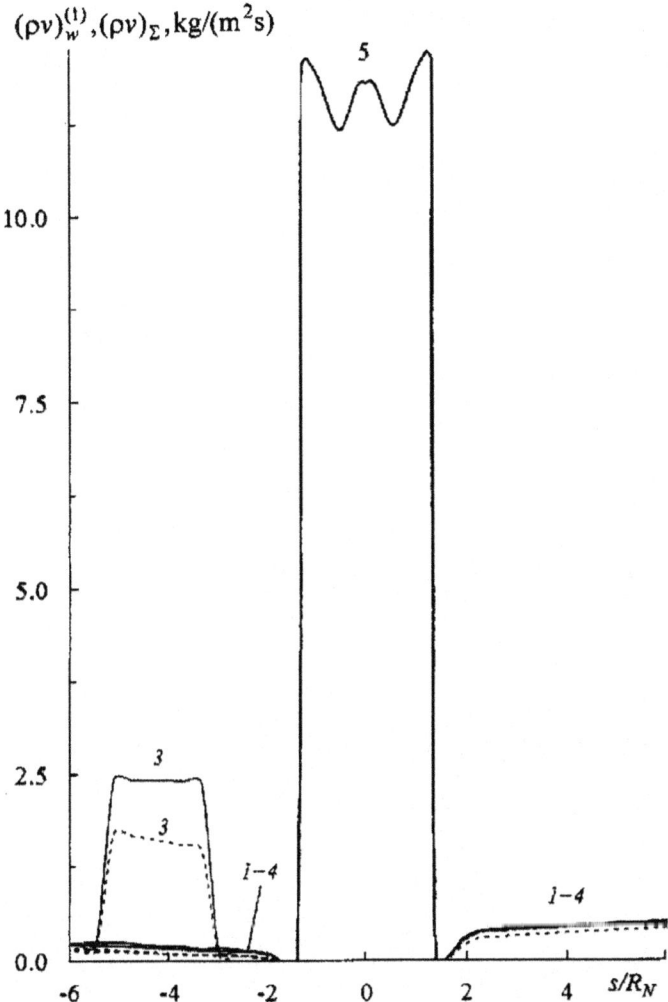

Fig. 3.13 Distribution of mass on the surface of the spherical ablation blunt cone in the plane of $\eta = 0–180°$ on the longitudinal coordinate on the windward and leeward side of the body. The notation is the same as in Fig. 3.12

internal wall temperature for carbon fiber reinforced plastic has not exceeded $T_{2l} = 293$ K, and for graphite without substrate at $L_0 = 0.01$ m we have $T_{2l} = 1730$ K at the moment $t = t_z$. For graphite with a substrate of asbestos cement (0.0025 m thick) at $L_0 = 0.01$ m, we obtain $T_{2l} = 896$ K, and at substrate thickness 0.005 m we have $T_{2l} = 352$ K.

The last result for graphite V-1 (the thermal conductivity of graphite V-1 is 25–30 times higher than the thermal conductivity carbon fiber reinforced plastic, their specific heat capacities differ by one and a half times [2, 124] in the considered

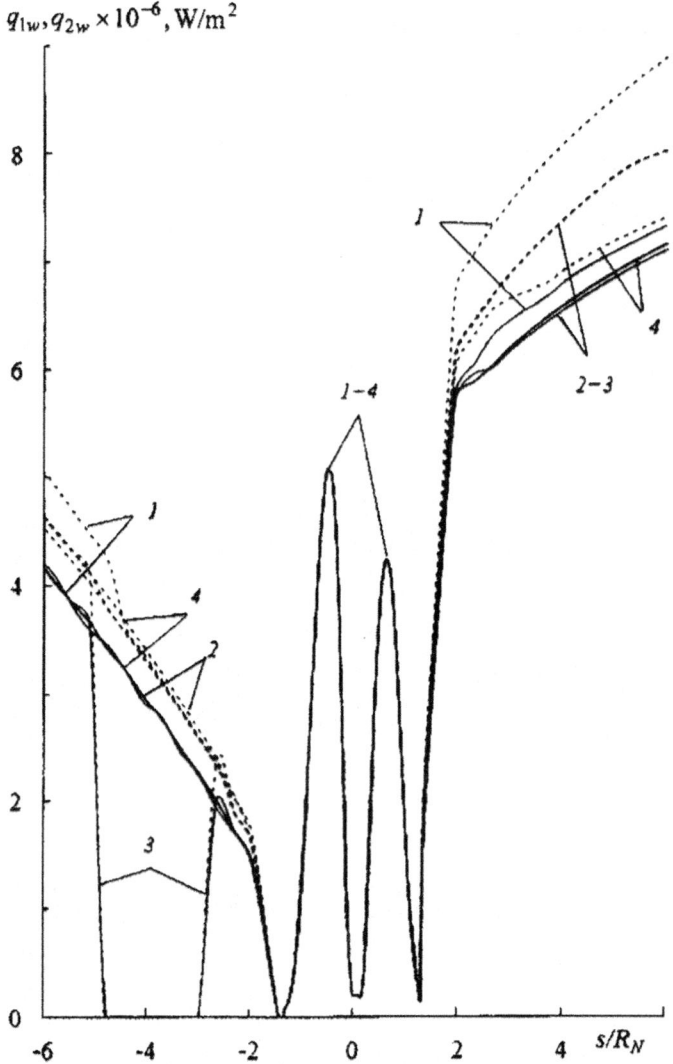

$q_{1w}, q_{2w} \times 10^{-6}, \mathrm{W/m}^2$

Fig. 3.14 Distribution convective heat flux distribution in the plane of $\eta = 0$–$180°$ longitudinal position. The notation is the same as in Fig. 3.12

temperature range) shows that to reduce the internal wall temperature it is necessary to use a composite material on the conical part of the body.

Neither that decreases twice in the formula for attenuated radiant component (3.2.1) at $z_1 = 1$, nor z_1 that increases twice in the latter equation (3.2.5) at $\xi = 1$ significantly inhibits the carbon fiber reinforced plastic heating process and leads to qualitative changes in the solution to the problem. At the same time, at $q_r = q_*$ ($\xi = 0$, $z_1 = 1$), the carbon fiber reinforced plastic surface temperature in the second

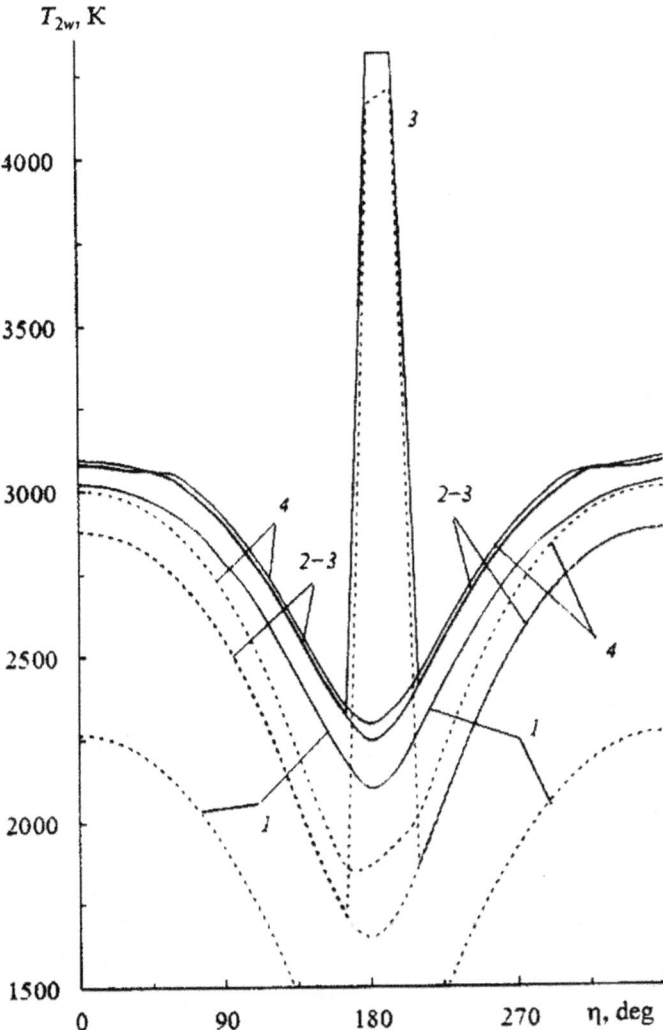

Fig. 3.15 Dependence temperature sectional surface $\bar{s}_1 = 3$ on the cone from the circumferential coordinate. The notation is the same as in Fig. 3.12

$t \geq t_4$ and third laser spikes will reach the maximum temperature T_{2w}: 4694–4703 K. This will lead to a higher heating rate of the composite material and more significant ablation of the material from the carbon fiber reinforced plastic surface (more than 22 times). It is obvious that such high values of heat mass and transfer characteristics cannot be implemented in practice due to mechanical carbon fiber reinforced plastic ablation and attenuation of the incident radiant flow with products of thermochemical decomposition.

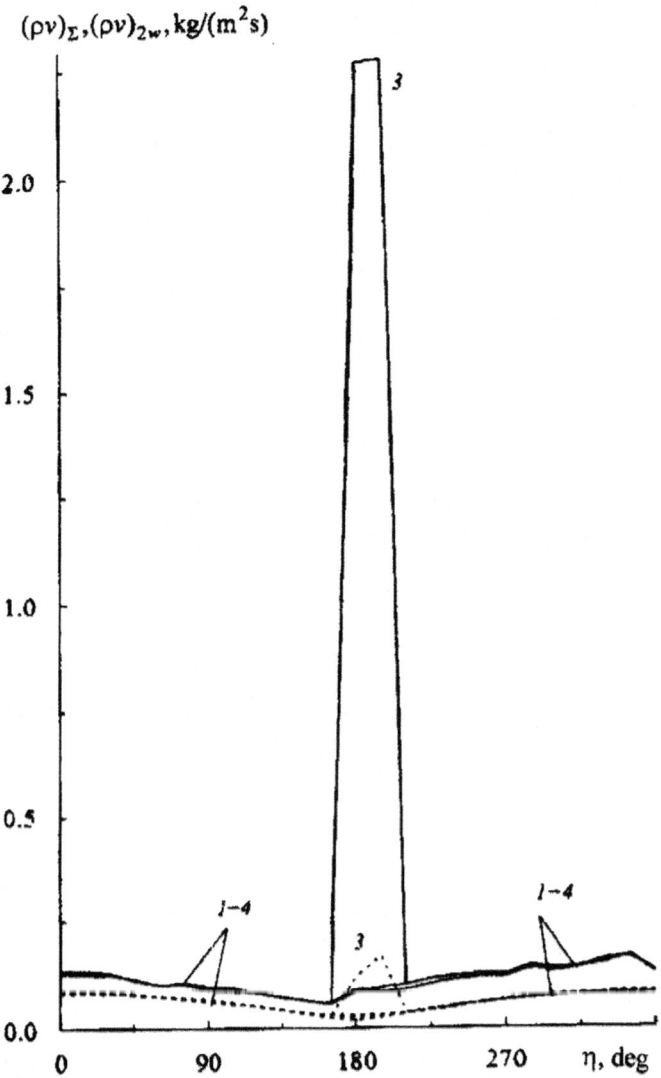

Fig. 3.16 Dependence of the total mass of ablation from the surface of carbon fiber for $\bar{s}_1 = 3$ from the circumferential coordinate. The notation is the same as in Fig. 3.12

The absence of convective heat in the total heat flow in action zones of the second and third laser spikes $(t_3 \leq t \leq t_6)$ does not change heat and mass transfer characteristics presented in Figs. 3.10, 3.11, 3.12, 3.13, 3.14, 3.15 and 3.16. The fact is that, at the time of radiation spikes, $(W = 1)$ q_r is more than an order of magnitude higher than q_{2w}, while, at $W = 0$, Q_Σ derived from (3.2.16) is largely determined by ablation of energy due to carbon fiber reinforced plastic thermo-chemical decomposition.

It should be noted that the formulas for densities of convective and diffusion flows q_{2w} and J_{iw} lose their physical meaning during strong thermochemical decomposition, as these formulas are based on concepts and methods of the theory of boundary layer, which is pushed away by strong injection of decomposition products. However, the limited nature of the formula for q_{2w} has almost no effect on the numerical results, since the value of the total heat flow Q_Σ during the exposure to radiation is largely determined by the radiant component. This conclusion agrees well with mathematical experimental data at $q_{2w} = 0$ presented above.

3.3 Conclusion

1. Vibrations of composite materials may stimulate interphase heat exchange between the binder and the filler, with additional vibration "transport" of heat into the depth of the material. This process is accompanied by earlier decomposition of the binder and filtration of gasification products through pores of the external surface. As a result, the temperature of the wall decreases. Due to the optimal choice of amplitude, vibration frequency and weight content of the binder, it is possible to regulate heat loads on composite materials and to control thermochemical destruction.
2. Based on the tested results [2, 49, 63, 65], the problem of the effect of surface roughness and ablation of coal-plastic thermochemical destruction products on heat transfer in thermal-protective materials was refined. Ablation through the rough surface, differently affects the rate of heat transfer in thermal-protective materials. The numerically calculated data were compared with the known data.
3. Based on the experimental [76, 77] and tested theoretical results [78, 80], a mathematical model for thermochemical destruction of the multilayer thermal-protective material was refined. By taking into account the flow across the body, the state of protected structures under fire conditions could be predicted more accurately. The lining layer made of intumescent materials is found to provide more reliable fire protection for the composite body. When the third layer (intumescent fire-retardant coating) in the multilayer structure is taken into account, there is less probability that the protected porous steel layer will be destroyed for a limited time of fire exposure ($t \leq 20$ min). This conclusion is consistent with the known data [80].
4. The mathematical model was developed to identify data of the experiment in transpiration cooling systems from permeable metals. Increasing thermal conductivity is found to decrease the temperature of the thermal-protective material, while increasing porosity results in more uniform distribution of the coolant over the surface and, therefore, a reduction of heat loads on the protected structure. The findings were proved to be consistent with the experimental data [2, 44] both in quantitative and qualitative terms.

5. Transpiration cooling systems are found to be sensitive to pulsations of injected gas and wall vibrations. Such sensitivity is explained by additional normal and tangential friction stresses arising when a coolant gas is filtered through pores, as well as by additional heat transfer into the depth of the porous wall. This makes it possible to regulate and control hydrodynamical and heat parameters of such systems to improve heat resistance, intensity of internal heat transfer, gas permeability of porous materials, and thermal protection effectiveness of transpiration cooling systems.

6. Based on laboratory experiment, a mathematical model was developed to calculate heat and mass transfer parameters in two-phase transpiration cooling systems with and without low-energy perturbations. Without low periodic perturbations, phase transition of a liquid in porous materials causes unstable cooling processes and wall temperature fluctuations. The latter factor must be adequately taken into account in the mathematical statement of the problem.

7. The conjugate formulation of the problem allows to consider the impact of non-isothermal shell wall on the characteristics of heat and mass transfer in the boundary layer.

   Investigated of rotation effects on parameters conjugate heat and mass exchange with a high-enthalpy flow around a spherically blunted cone at an angle of attack and ablation from the surface. The effects of rotational motion of a body on heat migration to a thermal-protective coating have been analyzed. It has been shown that the choice of thermal-protective material and optimal rotational speed makes it possible to control heat and mass transfer conditions when a body is moving at hypersonic speed.

8. Gaseous products of pyrolysis and condensed-phase particles are found to play a decisive role in shielding the body from laser radiation with carbon fiber-reinforced plastic decomposition products at the initial stage of its exposure to radiation. In case of repeated pulse loading, shielding for $T_{2w} > 3800$ K can be implemented mainly with carbon vapors and solid-phase particles.

Conclusion

1. Vibrations of composite materials may stimulate interphase heat exchange between the binder and the filler, with additional vibration "transport" of heat into the depth of the material. This process is accompanied by earlier decomposition of the binder and filtration of gasification products through pores of the external surface. As a result, the temperature of the wall decreases. Due to the optimal choice of amplitude, vibration frequency and weight content of the binder, it is possible to regulate heat loads on composite materials and to control thermochemical destruction.

2. Based on the tested results [2, 49, 63, 65], the problem of the effect of surface roughness and ablation of coal-plastic thermochemical destruction products on heat transfer in thermal protective materials was refined. Ablation through the rough surface, differently affects the rate of heat transfer in thermal protective materials. The numerically calculated data were compared with the known data.

3. Based on the experimental [76, 77] and tested theoretical results [78, 80], a mathematical model for thermochemical destruction of the multilayer thermal protective material was refined. By taking into account the flow across the body, the state of protected structures under fire conditions could be predicted more accurately. The lining layer made of intumescent materials is found to provide more reliable fire protection for the composite body. When the third layer (intumescent fire-retardant coating) in the multilayer structure is taken into account, there is less probability that the protected porous steel layer will be destroyed for a limited time of fire exposure ($t \leq 20$ min). This conclusion is consistent with the known data [80].

4. The mathematical model was developed to identify data of the experiment in transpiration cooling systems from permeable metals. Increasing thermal conductivity is found to decrease the temperature of the thermal protective material, while increasing porosity results in more uniform distribution of the coolant over the surface and, therefore, a reduction of heat loads on the protected structure. The findings were proved to be consistent with the experimental data [2, 44] both in quantitative and qualitative terms.

5. Transpiration cooling systems are found to be sensitive to pulsations of injected gas and wall vibrations. Such sensitivity is explained by additional normal and

© Springer International Publishing AG, part of Springer Nature 2018
A. S. Yakimov, *Thermal Protection Modeling of Hypersonic Flying Apparatus*,
Innovation and Discovery in Russian Science and Engineering,
https://doi.org/10.1007/978-3-319-78217-1

tangential friction stresses arising when a coolant gas is filtered through pores, as well as by additional heat transfer into the depth of the porous wall. This makes it possible to regulate and control hydrodynamical and heat parameters of such systems to improve heat resistance, intensity of internal heat transfer, gas permeability of porous materials, and thermal protection effectiveness of transpiration cooling systems.

6. Based on laboratory experiment, a mathematical model was developed to calculate heat and mass transfer parameters in two-phase transpiration cooling systems with and without low-energy perturbations. Without low periodic perturbations, phase transition of a liquid in porous materials causes unstable cooling processes and wall temperature fluctuations. The latter factor must be adequately taken into account in the mathematical statement of the problem.

7. The conjugate formulation of the problem allows to consider the impact of non-isothermal shell wall on the characteristics of heat and mass transfer in the boundary layer.

   Investigated of rotation effects on parameters conjugate heat and mass exchange with a high-enthalpy flow around a spherically blunted cone at an angle of attack and ablation from the surface. The effects of rotational motion of a body on heat migration to a thermal protective coating have been analyzed. It has been shown that the choice of thermal protective material and optimal rotational speed makes it possible to control heat and mass transfer conditions when a body is moving at hypersonic speed.

8. Gaseous products of pyrolysis and condensed-phase particles are found to play a decisive role in shielding the body from laser radiation with carbon fiber-reinforced plastic decomposition products at the initial stage of its exposure to radiation. In case of repeated pulse loading, shielding for $T_{2w} > 3800$ K can be implemented mainly with carbon vapors and solid-phase particles.

Bibliography

1. Shashkov AG. Heat and mass transfer in the heated gas flow. Minsk: Science and Engineering; 1974. p. 101.
2. Grishin AM, Golovanov AN, Zinchenko VI, Efimov KN, Yakimov AS. Mathematical and physical modeling of thermal protection. Tomsk: Tomsk University Publishing House; 2011. p. 358.
3. Avduevsky VS, Galitseysky BM, Glebov BA. Basic heat transfer concepts in aviation and aerospace engineering. Moscow: Mechanical Engineering; 1975. p. 624.
4. Polezhaev YV, Yurevich FP. Thermal protection. Moscow: Energy; 1976. 392p.
5. Gorsky VV. Theoretical basis for calculating ablation thermal protection. Moscow: Science World; 2015. p. 688.
6. Polezhaev YV. Current problems of thermal protection. Eng Phys J. 2001;74(6):8–16.
7. Polezhaev YV, Reznik SV, Vasilevsky EB, et al. Materials and coatings in extreme conditions. A look into the future: issue 3, vol. 1. In: Reznik SV, editor. Prediction and analysis of extreme conditions. Moscow: Publishing House of the N.E. Bauman Moscow State Technical University; 2002. 224 p.
8. Mikhatulin DS, Polezhaev YV, Reviznikov DL. Heat and mass transfer. Thermochemical and thermal erosion of thermal protection, Moscow: Yanus-K; 2011. 516 p.
9. Grishin AM, Fomin VM. Conjugate and nonstationary problems in the mechanics of reacting media. Novosibirsk: Science. Siberian Branch of the USSR Academy of Sciences; 1984. p. 319.
10. Nikitin PV. Thermal protection. Moscow: Publishing House of the Moscow Aviation Institute; 2006. p. 512.
11. Schneider PJ, Dolton TA, Reed GW. Mechanical erosion of charring Ablators in ground-test and re-entry environments. Rocket Space Technol. 1968;6(1):76–87.
12. April GC, del Valle EG, Pike RW. Modeling reacting gas flow in the char layer of an Ablator. Rocket Space Technol. 1971;9(6):148–56.
13. Zinchenko VI, Yakimov AS. Thermochemical destruction conditions of carbon phenolic composites under the effect of heat flow. Phys Combus Explos. 1988;24(2):141–9.
14. Grishin AM, Parashin AD, Yakimov AS. Thermochemical destruction of carbon fiber-reinforced plastics with repeated impulsive loading. Phys Combus Explos. 1993;29(1):87–95.
15. Sindyaev NI. Overview of research techniques for hypersonic flows around bodies with ablative coatings. Aeromech Thermo Phys. 2001;11(4):501–21.
16. Volchkov EP. Near-wall gas screens. Novosibirsk: Science; 1983. 239 p.
17. Leontyev AI, Volchkov EP, Lebedev VP, et al. Thermal protection of plasma gun walls, Novosibirsk: Institute of Thermophysics. Siberian Branch of the Russian Academy of Sciences; 1995. 336 p. (Low-Temperature Plasma, vol. 15).
18. Sidnyaev NI. Mathematical modeling of mass transfer for flows around hypersonic flight vehicles. Bull Moscow State Tech Univ Nat Sci Series. 2001;2(7):54–63.

© Springer International Publishing AG, part of Springer Nature 2018
A. S. Yakimov, *Thermal Protection Modeling of Hypersonic Flying Apparatus*,
Innovation and Discovery in Russian Science and Engineering,
https://doi.org/10.1007/978-3-319-78217-1

19. Vasilevsky EV. Injection-based thermal protection of the body surface against heat flow. Aeromech Gas Dyn. 2003;2:37–48.

20. Polyaev VM, Mayorov VA, Vasilyev LA. Hydrodynamics and heat transfer in porous elements of flight vehicle structures. Moscow: Mechanical Engineering; 1988. 168 p.

21. Golovanov AN. Heat transfer of the plasma jet and hemispherical wall with a coolant gas injected through round openings. Appl Mech Eng Phys. 1988;2:18–23.

22. Repukhov VM. Theory of wall thermal protection by gas injection. Kiev: Scientific Publications; 1980. p. 296.

23. Golovanov AN, Yakimov AS, Krasnov AS. Modeling of heat and mass transfer in transpiration cooling systems with phase transitions. High Temp. 2012;50(5):685–91.

24. Kutateladze SS, Nakoryakov VE. Heat and mass transfer and waves in gas-liquid systems. Novosibirsk: Science. Siberian Branch of the USSR Academy of Sciences; 1984. p. 301.

25. Zinchenko VI. Mathematical modeling of conjugate heat and mass transfer problems. Tomsk: Tomsk University Publishing House; 1985. p. 221.

26. Polyakov AF, Reviznikov DL. Numerical modeling of conjugate heat and mass exchange in case of convective and film cooling. High Temp. 1999;36(4):420–6.

27. Zinchenko VI, Yakimov AS. Research of heat mass characteristics with the flow around the spherically blunted cone under angle of attack and gas injection from the blunted surface. Appl Mech Eng Phys. 1999;4:162–9.

28. Zinchenko VI, Efimov KN, Yakimov AS. Research of conjugate heat and mass transfer characteristics in case of gas injection and thermochemical destruction of the body in flow. High Temp. 2007;45(5):749–55.

29. Zinchenko VI, Efimov KN, Yakimov AS. Calculation of conjugate heat- and mass transfer characteristics in case of spatial flow around the body using the combined thermal protection system. High Temp. 2011;49(1):81–91.

30. Golovanov AN, Rulyeva EV, Yakimov AS. Modeling of heat transfer in transpiration cooling systems with gas flow pulsations. High Temp. 2011;49(6):914–21.

31. Yakimov AS. Calculation of heat transfer characteristics in case of transpiration cooling. J Appl Mech Mater. 2015;756:366–71.

32. Bashkin VA, Reshetko SM. Maximal temperature of the blunted area with allowance for thermal conductivity. Proc Cent Ins Aero Hydrodyn. 1989;20(5):53–9.

33. Kuryachiy AP. Modeling of the combined thermal protection system of radiation evaporation design. High Temp. 1993;31(5):767–76.

34. Raushenbakh BV. Vibration combustion. Moscow: Main Publishing House of Physical and Mathematical Literature; 1961. p. 500.

35. Tretyakov PK. Management of pulsing combustion in high-temperature in ramjets. Phys Combust Explos. 2012;48(6):21–7.

36. Pakhomov MA, Terekhov VI. Effects of pulse frequency on heat transfer at the stagnation point of the impact turbulent jet. High Temp. 2013;51(2):287–93.

37. Kapitsa PL. Thermal conductivity and diffusion in a liquid medium under periodic flow conditions. J Exp Theor Phys. 1951;21(9):964–78.

38. Nakoryakov VE, Burdukov AP, Boldarev AM, Terleev PN. Heat and mass transfer in the sound field. Siberian Branch of the USSR Academy of Sciences. Novosibirsk: Institute of Thermophysics; 1970. 253 p.

39. Boreskov GK, Matros YS, Kiselev OV, Bunimovich GA. Implementation of the heterogeneous catalytic process under unsteady flow conditions. Proc USSR Acad Sci. 1977;237(1):160–3.

40. Galliulin RG, Repin VB, Khalitov NK. Viscous flow and heat transfer of bodies in the sound field. Kazan: Publishing House of the Kazan University; 1978. p. 128.

41. Azhishchev NL, Bykov VI. Intensification of heat transfer in porous media with pressure pulsation. In: Proceedings of the Siberian Branch of the USSR Academy of Sciences. Engineering science series 1987;6(21):27–30.

42. Markov AA. Effects of body rotation and external vorticity on heat transfer around the critical point of the blunted body in a hypersonic flow. Proc USSR Acad Sci Mech Fluids Gases 1984;3:179–82.

43. Bunkin FV, Kirichenko NA, Lukyanchuk BS. The thermochemical effect of laser radiation. Successes Phys Sci. 1982;138:(9):43–90.

44. Golovanov AN. Low-energy perturbations in some problems of the mechanics of reactive media. Tomsk: TML-Press; 2010. p. 118.

45. Grishin AM, Zinchenko VI, Efimov KN, Subbotin AN, Yakimov AS. The iterated interpolation method and its applications. Tomsk: Tomsk University Publishing House; 2004. p. 320.

46. Golovanov AN. Effects of periodic perturbations on thermochemical destruction of some composite materials. Phys Combust Explos. 1999;35(3):67–73.

47. Golovanov AN, Yakimov AS. Thermochemical destruction of carbon phenolic composites in the high-enthalpy pulsing gas flow. J Eng Phys. 2011;84(2):386–92.

48. Stepanova EV, Yakimov AS. mathematical modeling of heat and mass transfer in thermal protective coatings under gas flow pulsations. High Temp. 2015;53(2):236–42.

49. Krotov MK, Ovchinnikov VA, Yakimov AS. Mathematical modeling of influence surface roughness and ablation on thermal protection. In: 2015 international conference on mechanical engineering. Automation and control systems (MEACS). 2015 Dec 1–4. p. 1–5. https://doi.org/10.1109/meacs. 2015.7414933.

50. Ovchinnikova VA, Yakimov AS. Thermal protection of multi-layer structures under fire exposure. J Eng Phys. 2016;89(3):321–9.

51. Golovanov AN. Effects of vibrations on combustion of some coal-graphite materials. Phys Combust Explos. 1988;24(2):69–71.

52. Golovanov AN. Hydrodynamic and thermal characteristics of transpiration cooling systems in the presence of periodic low-energy perturbations. J Eng Phys. 1994;66(6):695–701.

53. Kalinkevich GA, Mikov VL, Morozova TP. Research of the polyaminimide binder by complex thermal analysis. Proc Timiryazev Agri Acad. 1981;2:164–7.

54. Gorsky VV, Zabarko DA, Olenicheva AA. Study of ablation of carbon materials within the framework of the complete thermochemical destruction model for the case of equilibrium chemical reactions in the boundary layer. High Temp. 2012;50(2):307–12.

55. Nikitin PV, Ovsyannikov VM, Kholodkov NV. Destruction of organic composite materials in the high-temperature gas flow. J Eng Phys. 1986;50(3):112–363.

56. Vasilevsky KK, Fyodorov OG. The study of internal heat transfer between gas and frame in ablating materials. Heat and mass transfer, vol. 2. Minsk: Science and Engineering 1968; p. 67–74.

57. Andrievsky RA. Porous metal-ceramic materials. Moscow: Metallurgy; 1964. p. 187.

58. Buchnev LM, Smyslov AI, Dmitriev IA, et al. The experimental study of enthalpy of graphite Quasi-monocrystal and glass carbon in the temperature interval of 300–3800 K. High Temp. 1987;2(6):1120–5.

59. Karapetyants MK, Karapetyants MM. Basic thermodynamic constants of inorganic and organic matters. Moscow: Chemistry; 1968. p. 471.

60. Vargaftik NB. Guide to thermophysical properties of gases and fluids. Moscow: Physical and Mathematical State Edition; 1963. p. 670.

61. Einstein A, Smoluchowski M. Brownian motion. Moscow-Leningrad: Main Publishing House of General Engineering Literature; 1936. p. 607.

62. Loitsyansky LG. Mechanics of fluids and gases. Moscow: Science. Main Publishing House of Physical and Mathematical Literature; 1973. p. 847.

63. Schlichting G. Boundary-layer theory. Moscow: Science. Main Publishing House of Physical and Mathematical Literature; 1969. p. 711.

64. Welsh WE. Shape and surface roughness effects on turbulent nose tip ablation. AIAA J. 1970;11:1983–189.

65. Nestler DE. Compressible turbulent boundary layer heat transfer to rough surfaces. AIAA J. 1971;9:1799–803.
66. Laganelli AL, Nestler DE. Surface ablation patterns: a phenomenology study. AIAA J. 1969;7:1319–25.
67. Droblenkov VF. The turbulent boundary layer on rough curved surfaces. In: Proceedings of the USSR Academy of Sciences. Department of Engineering Sciences 1955;8:17–21.
68. Golovanov AN, Zima VP, Stepanova EV. Thermal protection method for the aircraft head. Author's Certificate 2481239. Application No. 2012102950; 2012 Jan 27; published 2013 May 10; Bulletin No. 13, 7 p.
69. Zima VP, Stepanova EV. The study of thermal protection of flight vehicles using ablative composite materials. Important problems of thermal physics. Fluid and gas dynamics. In: Proceedings of the All-Russian conference and school with international participation Novosibirsk, on 2014 Nov 20–23; Novosibirsk: Institute of Thermophysics. Siberian Branch of the Russian Academy of Sciences; 2014. 195 p. Electronic version.
70. Baratov AN, Pchelintsev VD. Fire safety, Moscow: Publishing House, Association of Construction Institutes of Higher Education; 1997. 176 p.
71. Romanenkov IG, Levites FA. Fire protection of building structures. Moscow: Construction Publishing House; 1991. p. 320.
72. Intumescent Fire-Retardant Compound SGK-1/TU 7719-162-00000335-95. Moscow: Research and Production Enterprise "Spetsialnaya Energotekhnika"; 1995. 37 p.
73. Levites FA, Maryasin IA, Puklina OS. Modification of intumescent fire-retardant coating VPM-2. Fire-resistance of building structures. Moscow: All-Russian Research Institute for Fire Protection; 1988. 39 p.
74. Reshetnikov IS, Antonov AV, Khalturinsky NA. Mathematical description of combustion of intumescent polymer systems (overview). Phys Combust Explos. 1997;33(6):48–67.
75. Strakhov VL, Chubakov NG. Calculation of temperature fields in intumescent materials. J Eng Phys. 1983;45(3):472–9.
76. Isakov GN, Nesmelov VV. On some heat and mass transfer mechanisms in intumescent fire-retardant materials. Phys Combust Explos. 1994;30(2):57–63.
77. Isakov GN, Kuzin AJ. Modeling and identification of heat and mass transfer processes in intumescent fire-retardant materials. Appl Mech Eng Phys. 1996;37(4):126–34.
78. Zverev VG, Goldin VD, Nesmelov VV, Tsimbalyuk AF. Modeling of heat and mass transfer in intumescent fire-retardant materials. Phys Combust Explos. 1998;34(2):90–8.
79. Strakhov VL, Krutov AM, Davydkin NF. Fire protection of building structures, vol. 2. Moscow: Timr 2000. 433 p.
80. Zverev VG, Nazarenko VA, Tsimbalyuk AF. Fire protection of multilayer containers. High Temp. 2008;46(2):283–9.
81. Zinovyev VE. Thermophysical properties of metals at high temperatures handbook. Moscow: Metallurgy; 1989. p. 383.
82. Samarsky AA. Introduction to the theory of difference schemes, Moscow: Science; 1971. 552 p.
83. Sokolov PN. Technology of Asbestos-cement products. Moscow: Industrial Construction Publishing House; 1951. p. 352.
84. Vargaftik NB. Guide to thermophysical properties of gases and fluids. Moscow: Science; 1972. 720 p.
85. Hoff N. Introduction. Problems of high temperatures in aircraft structures. Moscow 1961:7–14.
86. Mayorov VA. Boundary values for intensive transpiration cooling systems. J Eng Phys. 1984;47(4):587–94.
87. Koh J, Colony R. Analysis of cooling effectiveness for porous material in a coolant passage. Heat transfer. Proc Am Soc Mech Eng Series C 1974;96(3):66–74.

88. Kubota H, Thermal characterization of an evaporative cooling system under conditions of combined radiative and convective heating. Heat transfer. Proc Am Soc Mech Eng Series C. 1977;99(4):132–8.

89. Dorot VL, Strelets MK. Transpiration cooling in a hypersonic turbulent boundary layer. High Temp. 1973;11(6):551–660.

90. Grishin AM, Laeva VI, Yakimov AS. Boundary conditions for the heat and mass transfer model of a two-temperature porous medium with gas flow. J Eng Phys. 1989;50(6):1029–30.

91. Ovchinnikov VA, Yakimov AS. Mathematical modeling of the heat exchange process transpiration cooling systems under influence of pulsations of a coolant gas. In: 2015 International conference on mechanical engineering. Automation and control systems (MEACS). 2015 Dec 1–4. 2015:1–6, https://doi.org/10.1109/meacs.2015.7414934.

92. Golovanov AN, Yakimov AS. Modeling of two-phase transpiration cooling in the presence of periodic low-energy perturbations. J Eng Phys. 2012;85(3):472–81.

93. Belov SV, Vityaz PA, Sheleg VK, et al. Porous permeable materials: handbook. Moscow: Metallurgy; 1987. p. 335.

94. Sovershenny VD. Engineering formulas for calculating friction on the permeable surface in the turbulent gas flow. J Eng Phys. 1967;12(4):538–43.

95. Sovershenny VD. Turbulent boundary layer on the permeable surface. Proc USSR Acad Sci Mech Fluids Gases 1966;3:45–51.

96. Bureev AV, Zinchenko VI. Calculation of spatial flow around spherically blunted cones in the vicinity of the symmetry plane under different flow conditions in the shock layer and gas injection from the surface. Appl Mech Eng Phys. 1991;6:72–9.

97. Grishin AM, Golovanov AN, Yakimov AS. Conjugate heat transfer in composite materials. Appl Mech Eng Phys. 1991;4:141–9.

98. Zanemonets VF, Rodionov VI. Experimental investigation of heat transfer in the granular bound layer. Heat and mass transfer. Minsk International Forum. Minsk: Institute of Heat and Mass Transfer, Academy of Sciences of the Byelorussian Soviet Socialist Republic 1988; Sec. 7:42–7.

99. Alifanov OM, Tryanin AP, Lozhkin AL. Experimental investigation of the method for determining the internal heat-transfer coefficient by solution of the inverse problem. J Eng Phys. 1987;52(3):461–9.

100. Golovanov AN, Ruleva EV. On the influence of periodic pulsations of the gas-cooler in the heat transfer characteristics of the system porous cooling. Bulletin of the Tomsk State University. Mathematics and Mechanics 2011;(2):85–90.

101. Mayorov VA, Vasilyev LL. Analytical investigation of resistance and heat transfer in two-phase cooling of a porous fuel element. Problems of heat and mass transfer. Minsk: Science and Engineering; 1976:232–258.

102. Pavlyukevich NV, Gorelik GE, Levdansky VV, et al. Physical kinetics and transfer processes in phase conversions. Minsk: Science and Engineering; 1980. p. 280.

103. Vukalovich MP, Rivkin SA, Aleksandrov AA. Tables of thermophysical properties of water and steam. Moscow: Publishing House of Standard; 1969. p. 408.

104. Zverev IN, Smirnov NN. Gas dynamics of combustion. Moscow: Publishing House of the Moscow State University; 1987. p. 308.

105. Golovanov AN. Generation of self-sustained oscillations by injecting a liquid into a plasma jet. J Eng Phys. 1991;61(4):650–5.

106. Krasilov NA, Levin VA, Yunitsky SA. Research of the hypersonic viscous layer on rotating bodies with injection. Proc USSR Acad Sci Mech Fluids Gases 1986;(1):106–14.

107. Kumari M, Nath G. Heat and mass transfer unsteady compressible axisymmetric stagnation point boundary layer flow over a rotating body. Int J Heat Mass Trans. 1982;25(2):290–3.

108. Sharma BI. Computation of flow past a rotating cylinder with an energy-dissipation model of turbulence. AIAA J. 1977;15(2):271–4.

109. Koosinlin MI, Lockwood FC. The prediction of axisymmetric turbulent swirling boundary layers. AIAA J. 1974;12(4):547–54.

110. Khoskin NE. A laminar boundary layer on a rotating sphere. Boundary layer problem and issues of heat and transfer. Leningrad. Publishing House "Energy" 1960;(2):114–8.

111. Gershbein EA, Peygin SV. The hypersonic viscous impact layer in the vortex flow on the permeable surface. Proc USSR Acad Sci Mech Fluids Gases 1986;(6):27-37.

112. Zinchenko VI, Kataev AG, Yakimov AS. Research of temperature conditions of bodies in flow with gas injection from the surface. Appl Mech Eng Phys. 1992;6:57–64.

113. Efimov KN, Ovchinnikov VA, Yakimov AS. Numerical study of the characteristics of the conjugate heat and mass transfer in a hypersonic flow around a rotating spherical spatial blunt body and blowing gas from the surface. 15-th Minsk International Forum on 2016 May 23–26. Abstracts. Minsk: Publishing House IHME name AV. Lykov NAN Belarus; 2016. 4 p.

114. Grishin AM, Zinchenko VI. Conjugate heat transfer between the reactive solid body and gas in the presence of non-equilibrium chemical reactions. Proc USSR Acad Sci Mech Fluids Gases 1974;(2):121-128.

115. Lunev VV, Magomedov KM, Pavlov VG. Hypersonic flow around blunted cones with allowance for equilibrium physical and chemical transitions. Moscow: Computing Center of the USSR Academy of Sciences; 1968. p. 203.

116. Patankar SV, Spalding DB. Heat and mass transfer in boundary layers Moscow: Energy; 1970. 127 p.

117. Cebeci T. Behavior of turbulent flow near a porous wall with pressure gradient. AIAA J. 1970;8(12):48–52.

118. Kovalev VL. Heterogeneous catalytic processes in aerothermodynamics. Moscow: Physical and Mathematical Literature Publishing House; 2002. p. 224.

119. Feldhuhn RN. Heat transfer from a turbulent boundary layer on a porous hemisphere. New York. 1976. (Pap/AIAA; No. 119).

120. Widhopf. Hall turbulent heat and transfer measurements on a blunted cone at angle of attack under transition and turbulent flow conditions rocket and space technology 1972;10 (10):71–9.

121. Gofman AG, Grishin AM. Theoretical research of thermochemical destruction of graphite in high-enthalpy air. Appl Mech Eng Phys. 1984;4:107–14.

122. Baker R. Graphite sublimation chemistry nonequilibrium effects. Rocket Space Technol. 1977;15(10):21–9.

123. Kurshin AP. Hydrodynamic characteristics of permeable cermet. Proc Cent Thermo Gasdyn Ins. 1984;2230:10–42.

124. Sosedov VP. Properties of structural materials based on carbon. Directory. Moscow: Metallurgy, 1975. 335 p.

125. Vilenskaya GG, Nemchinov IV. The burst of laser radiation absorption and associated gas dynamic effects. Proc USSR Acad Sci. 1969;186(5):1048–51.

126. Anisimov SI, Imas YA, Romanov GS, Khodyko YV. The effect of high-power radiation on metals, Moscow: Science; 1970. 212 p.

127. Kirichenko NA, Korepanov AG, Lukyanchuk BS. On the change in the screening effect of thermal decomposition products of materials under the action of laser radiation in a moving medium. Quantum Electron. 1980;7(9):2049–51.

128. Minko LY, Goncharov VK, Loparev AN. The study of destructive effects of laser radiation reflection on opaque dielectrics. Phys Chem Mater Process. 1979;1:31–6.

129. Loparev AN, Minko LY. The role of particles in shielding effects of internal laser Erosion Plasma Jets. Phys Chem Mater Process. 1979;1:31–6.

130. Klimkov YM, Mayorov VS, Khoroshev MV. Interaction of laser radiation with the matter: textbook. Moscow: Moscow State University of Geodesy and Cartography; 2014. p. 108.

131. Libenson MN, Yakovlev EB, Shandybina GD. Interaction of laser radiation with the matter (Power optics). In: Veyko VP, editor, Part II. Laser heating and destruction of materials. Textbook. Saint Petersburg: ITMO University; 2014. 181 p.

132. Gorsky VV, Evdokimov IM, Zaprivoda AV, Resh VG. Attenuation of the Radiation Heat Flux with Vapors of Materials During Laser Cutting of Fiberglass. High Temp Thermophys. 2014;52(1):126–30.

133. Tsvetkov VB, Tsarkova OG, Garnov SV, et al. Modeling of temperature fields in flat targets under the influence of intense laser radiation. Proc Prokhorov General Phys Ins Russian Acad Sci. 2014;70:83–90.

134. Levdansky VV, Leytsina VG, Martynenko OG, Pavlyukevich NV. Radiation heating of the porous body [in Russian]. In: Rykalin NN, editor, The impact of concentrated energy fluxes on materials. Moscow: Scienceж 1985:99–107.

135. Zakharov NS, Karpenko VA, Shentsov NI. Interaction of optical radiation with structurally heterogeneous dielectrics [in Russian] High-temperature Thermophysics 1989;27(6): 1174–8.

136. Scala SM, Gilbert LM. Sublimation of graphite at hypersonic speeds. Rocket Space Technol. 1965;3(9):87–126.

137. Rykalin NN, Uglov AA, Zuev IV, Kokora AN. Laser and electron beam processing of materials. handbook. Moscow: Mechanical Engineering, 1985. 496 p.

138. Zemlyansky BA, Stepanov GI. Calculation of heat transfer parameters for a hypersonic air flow around thin blunted cones. Bull Acad Sci USSR Mech Fluids Gases 1981;5:173–7.

139. Charchenko VN. Heat transfer in a hypersonic turbulent boundary layer with injection of cooling gas through a gap. High Temp Thermophys. 1972;10(1):101–5.

140. Toshiyuki S, Kazuhisa F, Keisuke A, Takeharu S. Experimental study of graphite ablation in nitrogen flow. J Thermophys Heat Trans. 2008;22(3):382–9.

141. Gurvich LV, Hachkuruzova GA, Weits IV, Medvedeva VA. Thermodynamic properties of individual substances. Under the editorship of academician Glushko, VP. Moscow: Publishing House of the USSR Academy of Sciences; 1962(2):916 p.

Index

A
Active thermal protection, 31

B
Balance boundary conditions, 35, 56

C
Carbon fiber, 1, 2, 73, 74, 82, 84, 88, 89, 92, 94, 97, 98, 100
Coked layer, 8, 14, 94
Composite polymeric materials, 1, 16, 23, 28, 68, 81, 82, 84, 98, 101, 103
Conjugate heat, 68, 102, 104

D
Dispersion of carbon particles, 1

F
Flow around the body at an angle of attack, 67

H
Heat and mass transfer, 2, 3, 21, 22, 31, 32, 42, 46, 49, 51, 60, 72, 84, 95, 102, 104
Heat transfer due to body rotation, 5, 8, 13, 14, 16, 32, 36, 43, 44, 48, 51, 55, 61, 81, 102, 104

M
Mass entrainment of gaseous filtration products, 8, 13, 14

M
Mass transfer, 1, 4, 14, 22, 32, 42, 46, 68, 72, 82, 89, 95
Mass transfer in a two-temperature permeable medium, 49
Modeling of heat, 13, 32, 49, 82

N
Non-deformable porous reaction body, 1

P
Porous inert metal materials, 32
Pulsations of the gas stream, 2, 43, 65

R
Rotation around the longitudinal axis, 67

S
Screening of laser radiation, 82, 83
Small energy disturbances, 49
Surface roughness, 2, 16, 101, 103

T
Thermochemical destruction, 1, 2, 5, 7, 15, 29, 101, 103
Thermochemical destruction of carbon fiber, 2
Transpiration cooling, 31, 32, 42, 49, 52, 101–104
Two-phase porous cooling, 60

© Springer International Publishing AG, part of Springer Nature 2018
A. S. Yakimov, *Thermal Protection Modeling of Hypersonic Flying Apparatus*,
Innovation and Discovery in Russian Science and Engineering,
https://doi.org/10.1007/978-3-319-78217-1

Printed by Printforce, the Netherlands